# Climate Science Concepts Born in Hamburg

Hans von Storch • Martin Claussen •
Martin Heimann • Robert Sausen • Eduardo Zorita

# Climate Science Concepts Born in Hamburg

Springer

Hans von Storch
Hamburg, Hamburg, Germany

Martin Heimann
Max Planck Institute for Biogeochemistry
Jena, Thüringen, Germany

Eduardo Zorita
Helmholtz-Zentrum Hereon
Geesthacht, Schleswig-Holstein, Germany

Martin Claussen
Max Planck Institute for Meteorology
Hamburg, Germany

Robert Sausen
Institut für Physik der Atmosphäre
Deutsches Zentrum für Luft & Raumfahrt
Oberpfaffenhofen, Bayern, Germany

ISBN 978-3-031-81107-4 ISBN 978-3-031-81108-1 (eBook)
https://doi.org/10.1007/978-3-031-81108-1



Cover illustration: Air photo of Hamburg, taken by Dr. Andreas Villwock, with the TV tower and, in the red circle, the Geomatikum, for many years home of MPI and Meteorological Institute of University of Hamburg.

This Springer imprint is published by the registered company Springer Nature Switzerland AG
The registered company address is: Gewerbestrasse 11, 6330 Cham, Switzerland

# Foreword

The 1970s were an exciting time for atmospheric and climate sciences. In Europe, scientific and technological cooperation led to the establishment of the European Centre for Medium-Range Weather Forecasting (ECMWF), which has since been the world leader in numerical weather prediction (NWP). In the same year, 1975, the Max Planck Institute for Meteorology (MPI-M) was founded in Hamburg with Klaus Hasselmann as a founding director. Over the next 40 years, which are covered by this book, climate research based on numerical models advanced profoundly, evolving from simple box and column models to sophisticated global simulations representing atmospheric, land, and ocean processes, including the sources, sinks, and dynamics of carbon and other constituents not typically considered by NWP models, which focus on time scales from days to weeks.

Fifty years after the foundation of MPI-M, global climate modeling has reached a resolution where the effects of increasing greenhouse gas emissions and land use on physical processes across many scales from the surface to the mesosphere are well resolved. Notably, the climate modeling and NWP modeling communities have increasingly merged their efforts, collaborating on "digital twin" projects aimed at producing coupled atmosphere-land-ocean simulations to support adaptation to a warming climate and the resulting increase in extreme events. This remarkable progress supports the quantification of the impacts of various factors within the climate system, advancing well beyond the "detection and attribution" concepts envisioned by Klaus Hasselmann 50 years ago.

However, our confidence in climate predictions continues to rely heavily on theoretical frameworks and simplified models, which offer mechanistic descriptions of processes and scale interactions, providing crucial causal explanations for atmospheric and oceanic behavior.

In this book, five esteemed authors and colleagues—Hans von Storch, Martin Claussen, Martin Heimann, Robert Sausen, and Eduardo Zorita—demonstrate how their and their colleagues' research in Hamburg successfully applied problem reduction to enhance our understanding of the Earth's climate system. They investigated not only recent few centuries, during which human influence on the climate has been substantial but also provided insight into admissible solutions from much earlier in

Earth's history (and possibly beyond our own time). The authors show how the origins and effects of climate change were first identified through observations, physical principles, and mathematical formulations, then transformed into computer models to capture nonlinear interactions, coupling land, atmosphere, and ocean in physically meaningful ways. Finally, they interpreted modeling results through theoretical concepts applied to diagnose and evaluate the simulated processes.

It has been a great pleasure and an honor to host the preparational seminar series on aspects of climate research developed in Hamburg. I am confident that this book will attract readers among the next generation of climate scientists, eager to learn how great ideas and research are born. Regardless of what the future holds, including the increasing reliance on machine learning to derive predictions learning from imperfect models, the need for new ideas and theoretical frameworks to interpret real-world phenomena within model outputs will remain essential. In this regard, the descriptions of key building blocks of climate science presented in this book will undoubtedly continue to be relevant for many years to come.

Professor of Theoretical Meteorology at the University of Hamburg, Hamburg, Germany
September 2024

Nedjeljka Žagar

# Preface

## Background and Motivation

Since the foundation of the Max Planck Institute for Meteorology in 1975, climate science in Hamburg has seen a remarkable boost. Various ideas were brought forward, implemented, and tested. Many of them ignited interest in the global scientific community, thus adding significant momentum to the development of modern climate science. In many cases, the now-Nobel laureate Klaus Hasselmann was the source of the ideas, but the special atmosphere created by him motivated others in Hamburg to pursue new ideas, which eventually caught the interest of the global community.

It was Nedjeljka Žagar of the Meteorological Institute of the Universität Hamburg who initiated a seminar with the title "Climate Science Born in Hamburg." The contributors to this book have presented and tested their ideas and views on this topic in that seminar in 2022 and 2023. The presentations addressed innovative concepts and ideas in climate system science and modelling, mainly from the 1970s to the 1990s. Hence the focus of the seminar—and of this book—is on the dynamics of the global climate system and the interaction between its subsystems, rather than on meteorological or climate processes or climatology in general. And it is about scientific ideas, not about enabling science by national and international organizations, to which Hamburg scientists have contributed significantly.

Obviously and unavoidably, the selection of topics is, to some extent, subjective, and another group of authors may have arrived at a different set of topics. Ours are chosen to describe ideas that have been suggested by scientists while working in Hamburg and have been taken up by the international community in further applications and advancements. As is often the case, such ideas may have been presented by others before, but their publication in Hamburg caused a response in the community.

Our selection of topics are:

- Chapter 1 gives a very brief summary of the history of climate science in Hamburg to sketch the local scientific background and environment in which climate system science in Hamburg developed. A comprehensive overview of the history of climate science in Hamburg or the foundation of the Max Planck Institute for Meteorology has been published elsewhere (e.g., von Storch et al. 2018; Rosenhagen et al. 2010; Wax 2020).

- Chapter 2 discusses Klaus Hasselmann's concept of *noise and signal*, with the implications of the emergence of internally generated variability on large and long scales (stochastic climate model), the need to separate effect and noise in numerical experimentation, and an approach to clarify the presence of the impact of changing external drivers (*detection and attribution of climate change*). The broad spectrum of Hasselmann's ideas and impact on science, ranging from turbulence theory to climate variability to climate-society interaction, is discussed in another book (von Storch 2022).
- Chapter 3 presents the accounting for the *carbon cycle in quasi-realistic climate models*, next to atmospheric and oceanic dynamics and the development of techniques to constrain these using comprehensive atmospheric observations.
- Chapter 4 discusses the search for *multiple equilibria and abrupt changes in the global climate system* using quasi-realistic climate system models. Such flip-flop transitions were found in the 1960s and 1970s in climate models with a high degree of idealization, and the question arises whether such abrupt transitions would only be the result of oversimplified models.
- Chapter 5 comprises three concepts and *contributions to modelling techniques* that eventually enabled comprehensive atmosphere-ocean models: The numerical modelling by Walter Hansen at the Institute of Oceanography of the University of Hamburg opened the scene for this field, which was at that time rather new technique (Sect. 5.1). Ernst Maier-Reimer constructed a computationally very efficient ocean model suitable for long ocean simulations on the limited computing resources of that time. The idea of this model is based on brilliant and consequent exploitation of the quasi-geostrophic approximation (Sect. 5.2). Finally, the *anomaly-coupling* approach to constrain climate models, which couple atmospheric and oceanic dynamics, allowed the construction of atmosphere-ocean models with stationary statistics and more reliable results from climate change simulations (Sect. 5.3).
- Chapter 6 introduces the *downscaling* concept, both with its empirical and its dynamical variants, which describes regional or local statistics as conditioned by large scales, and allows for the reconstruction of past regional variability and the regionalization of global climate change projections. We first sketch the concept of scales and how we may determine and separate climate variability of different spatial scales. A relevant detail is that climate models are mostly skilful on large scales, but less so on small scales. Downscaling links sufficiently well described large scales simulated in climate models, or represented in analyses, to statistics on smaller scales. Such links may be determined empirically or by means of dynamical models, both regional and global.
- Chapter 7 is an account on *paleoclimate and historical climate reconstructions*, it presents four important contributions to paleoclimatology and historical climate that arose within the Hamburg climate network. Three of them analyse available

indirect observations to infer past climate variations in the recent past and also at longer glacial timescales. The fourth contribution is more methodological and is focused on ways to test statistical climate reconstruction methods using synthetic data.

A responsible author was assigned to each of the chapters of this book, but all authors have joint responsibility for the whole book. That is, prior to submission of the manuscript, consensus has been achieved among the authors about the content. These responsible lead authors are Hans von Storch (Chaps. 2 and 6), Martin Claussen (Chaps. 1 and 4), Martin Heimann (Chap. 3), Robert Sausen (Chap. 5), and Eduardo Zorita (Chap. 7). The overall head of the project was Hans von Storch.

Unfortunately, we could not enrol Ernst Maier-Reimer in our project after he passed away in 2013. He was one of the key enablers of the progress made at the Max-Planck-Institute for Meteorology with his open mind, rigour, patience, and ability to know everything but say little.

Hamburg, Germany — Hans von Storch
Hamburg, Germany — Martin Claussen
Jena, Germany — Martin Heimann
Oberpfaffenhofen, Germany — Robert Sausen
Geesthacht, Germany — Eduardo Zorita
September 2024

## References

Rosenhagen G, Schatzmann M, Schrön A (2010) Das Klima der Metropolregion auf Grundlage meteorologischer Messungen und Beobachtungen. In: von Storch H, Claussen M (eds) Klimabericht für die Metropolregion Hamburg. Springer, Berlin, pp 19–59. https://doi.org/10.1007/978-3-642-16035-6

von Storch H, Claussen M, Gräbel C (2018) Zur Geschichte der Klimaforschung in Hamburg. In: Nicolaysen (ed) Geschichte U Hamburg, in press

von Storch (ed) H (2022) From Decoding Turbulence to Unveiling the Fingerprint of Climate Change: The Science of Klaus Hasselmann. Springer Publisher

Wax G (2020) Wissenschaft zwischen Planung, Aufgabenverteilung und Kooperation. Zum Aufstieg der Erdsystemforschung in der Max-Planck-Gesellschaft, 1968-2000, Preprint 5 des Forschungsprogramms Geschichte der Max-Planck-Gesellschaft. Max-Planck-Gesellschaft. https://doi.org/10.17617/2.3224614

# Obituary

We are deeply saddened and shocked to hear the news about the tragic loss of Robert Sausen, right after completing his contributions to this volume. We believe the scientific community has lost a world leading researcher. Our most sincere condolences!

# Acknowledgements

The authors are grateful to many colleagues: Hartmut Heinrich and Oliver Krüger checked the summary of the Heinrich-event and the Heiner-Schmidt story, Jürgen Sündermann helped to grasp the achievements of Walter Hansen (Sect. 5.1). The achievements of Ernst Maier-Reimer in numerical modelling the ocean and its carbon component (Sects. 5.2 and 3.6) have been prepared with help from Christoph Heinze. Michael Böttinger of DKRZ provided the diagrams illustrating the steadily improving grid resolutions of climate models (Figs. 6.5 and 6.8). We thank Stephan Bakan, Armineh Barkordarian, Richard Blender, Frauke Feser, Gabriele Hegerl, René Laprise, Frank Lunkeit, Thierry Penduff, and Sönke Zaehle (in alphabetic order) for discussion, providing us with various details, figures, and checking parts of the manuscript. The cover photo, captured by Andreas Villwock during a flight over Hamburg at our request, features the Geomatikum prominently at the center.

# Contents

# Chapter 1
# A Brief History of Climate Science in Hamburg

**Martin Claussen, Hans von Storch, Martin Heimann, Robert Sausen, and Eduardo Zorita**

**Abstract** This chapter provides a very brief summary of the history of climate science in Hamburg to sketch the local scientific background and environment in which climate system science in Hamburg developed.

With the ability to measure the weather, interest in the topic of climate seems to have initially focused on the local time series and derived statistics of temperatures, precipitation, and other weather variables. Also in Hamburg, regular weather records were kept by private individuals. In 1826, a first description of the climate and weather of Hamburg appeared by the physician Heinrich Wilhelm Buek on the basis of the series of observations from 1807–1826 by the Senior[1] Dr. Willerding Buek (1826). Many years of observations, from 1832–1873 and 1856–1886, respectively, were then followed by Zimmermann and the high school professor Scharenberg (Rosenhagen et al. 2010).

When the Norddeutsche Seewarte (North German Naval Observatory) was founded in 1866 and began its work two years later, observations and analyses of the climate took on a new quality, characterised in particular by a global perspective, instrumentation and advice for maritime shipping. After the foundation of the German Empire in 1871, the Norddeutsche Seewarte became the Deutsche Seewarte (German Naval Observatory), which advanced to the status of an imperial institute in 1875. The landmark building (Fig. 1.1) was physically destroyed in April 1945, the last month of World War II in Germany. After 1945, the Deutsche Seewarte was divided into a hydrographic service (Deutsches Hydrographisches Institut; DHI, today Bundesamt für Seeschifffahrt und Hydrographie, BSH) and an atmospheric unit assigned to the Deutscher Wetterdienst, the Seewetteramt.

The focus of the Deutsche Seewarte was the maritime meteorological and climatological service. Germany was at that time, the end of the nineteenth and the beginning of the twentieth century, acquiring colonies in Africa, Asia and Oceania, and the Deutsche Seewarte was responsible not only for running meteorological

[1] Leading clergyman of the Lutheran church in Hamburg.

 
H. von Storch et al., *Climate Science Concepts Born in Hamburg*,
https://doi.org/10.1007/978-3-031-81108-1_1

**Fig. 1.1** The Seewarte at Landungsbrücken in Hamburg, where prominent climate scientists such as Wladimir Köppen, Eduard Brückner and others worked. The building was destroyed in the last month of World War II (The picture is a postcard from 1906, coloured by Dario Ružić, with permission.)

monitoring in these colonies, but also for making data available for planning colonial acquisitions.

As the most important climatologist of Hamburg in historical times, the then already famous *Meteorologe der Deutschen Seewarte* Wladimir Köppen (1846–1940; see Fig. 1.2) is usually mentioned. Köppen worked at the Deutsche Seewarte from 1875 to 1919. Other famous and well-known scientists are Alfred Wegener (1880–1930), Köppen's successor as Meteorologe der Deutschen Seewarte (and son-in-law), as well as Eduard Brückner (1862–1927) and Georg von Neumayer (1826–1909). Wladimir Köppen was appointed by Georg von Neumayer as head of department III for weather telegraphy and storm warning and established - for the first time in Germany - a synoptic weather service. Köppen is mostly recognized for his *climate formula* in which he consistently defined climatic zones in relation to the geographic distribution of large-scale vegetation, or ecosystem, pattern. His global maps of climate zones are still used in climate studies and in many, if not most, school books today. And under the lead and initiative of Hugo Hildebrand Hildebrandsson, professor of meteorology in Uppsala, Sweden, Köppen and von Neumayer published the first international, multilingual cloud atlas (Hildebrandsson et al. 1890).

Wladimir Köppen also initiated and operated a kite station in Großborstel[2] from 1903, with a focus on aerological research work. In 1921, under the direction of

[2] www.grossborstel.de/wladimir-koeppen-und-die-drachenstation/.

**Fig. 1.2** Two towering climate scientists from Hamburg: Wladimir Köppen (1846–1940) and Klaus Hasselmann (1932–). Photo credit: Nobel Prize Outreach, Bernhard Ludewig

Alfred Wegener, the station was expanded into the Meteorological Experimental Station, where observation and calibration procedures for the meteorological instruments in use were developed.

Together with Wegener, Köppen published the book *Die Klimate der Geologischen Vorzeit* (The Climates of the Geological Past) in 1924. In this groundbreaking book, Köppen and Wegener invited the Serbian mathematician Milutin Milankovitch to present his quantitative calculation of the timing and duration of the European glacial–interglacial cycles, or *ice ages*. These Milankovitch curves are still considered a major landmark in the history of paleoclimate modelling.

Another important researcher at the Deutsche Seewarte was Erich Kuhlbrodt (1891–1972), who habilitated at the Universität Hamburg in 1923 and who was appointed associate professor at the Universität Hamburg in 1928 - like Alfred Wegener in 1919. He is known for his research on the general circulation of the atmosphere, especially for his high-altitude wind measurements aboard the research vessel *Meteor* during the German Atlantic Expedition from 1925 to 1927.

In 1919, the Universität Hamburg was founded, built upon the Hamburgisches Kolonialinstitut. Climate was studied in the Faculty of Geography. A chair in meteorology, established in 1929, did not significantly contribute to climate research until the 1970s. Until 1945, the Universität Hamburg was struggling with the consequences of the ideology of National Socialism, which quite a few university members, including those from meteorology and climatology, had adopted.

Climate science in Hamburg entered a kind of hiatus until 1975, when the Max Planck Society (Max-Planck-Gesellschaft, MPG) decided to build an institute for climate science, the Max-Planck-Institut für Meteorologie (MPI-M). The discussion about the foundation of the MPI-M took place at a time of growing environmental awareness in the late 1960s and early 1970s. At the MPG, Christian Junge, director at the MPI for Chemistry in Mainz from 1968 to 1979, promoted the idea of integrative atmospheric research, now known as Earth system research. The decisive initiative to found the MPI-M came with the offer from the German Federal Ministry for Research and Technology (in 1994 renamed to Federal Ministry for Education and Research) to the MPG to take over the internationally renowned Fraunhofer Institute for Radiometeorology and Maritime Meteorology, until 1972 lead by Karl Brocks, professor of meteorology at the Universität Hamburg. The main focus of the newly established MPI-M was to study the physical and chemical processes in the atmosphere and their interaction with the ocean and the biosphere. The name of the institute, MPI for Meteorology, was chosen for lack of a better term, as pointed out by the founding director of the MPI-M Klaus Hasselmann (Fig. 1.2), in his inauguration address. The MPI-M was first located in a university building, the Geomatikum, before it spread to adjacent buildings (Fig. 1.3). (For a comprehensive historical account of the rise of Earth system science in the MPG and the foundation of the MPI-M, in particular, see Wax (2020)).

When Hasselmann was appointed director of the MPI-M, he was known as a physicist with a reputation for ocean wave dynamics and modelling. He and his coworkers, Peter Lemke, Dirk Olbers, Claude Frankignoul, Jürgen Willebrand, and others, laid the theoretical foundations for a modern, empirical and theoretical climate science (von Storch and Heimbach 2022). At this time, the *Stochastic Climate Model* was designed. It was crucial to deal with the complex, nonlinear and high-dimensional climate system, which features both internal variability and externally forced changes (see Chap. 2). It was this achievement that earned Klaus Hasselmann his Nobel Prize in Physics more than 40 years later, in 2021: *Klaus Hasselmann created a model that links together weather and climate, thus answering the question of why climate models can be reliable despite weather being changeable and chaotic. He also developed methods for identifying specific signals, fingerprints, that both natural phenomena and human activities imprint in the climate. His methods have been used to prove that the increased temperature in the atmosphere is due to human emissions of carbon dioxide.*[3]

Reimar Lüst (1923–2020, Fig. 1.4) was the president of the MPG who founded the MPI-M. He was a German astrophysicist, who sought advice from Hermann Flohn, professor of meteorology and climatology at the Universität Bonn and Bert Bolin, professor at the Meteorological Institute of the Stockholm University, who was later, 1988, one of the founders of the Intergovernmental Panel on Climate

[3] https://www.nobelprize.org/prizes/physics/2021/press-release/.

**Fig. 1.3** The building Bundesstr. 53 in Hamburg-Rotherbaum in which the Max Planck Institute for Meteorology and the Institute for Oceanography of the Universtät Hamburg are housed. In the background, the *Geomatikum* where on the 17th floor, the MPI-M and its laboratories started in 1975, and where until the 2020s geosciences and climate science institutes of the Universität Hamburg are located (see https://de.wikipedia.org/wiki/Max-Planck-Institut für Meteorologie, Photo credit: Minderbinder under CC-BY-SA 4.0)

**Fig. 1.4** Reimar Lüst (right): As president of the Max-Planck Society, he decided to found a Max Planck Institute for Meteorology in Hamburg, with Klaus Hasselmann (left) as director

Change (IPCC). Reimar Lüst was an influential president of the MPG and later director general of the European Space Agency. Many years later, in 1990 after his retirement at ESA, Lüst became president of the Alexander-von-Humboldt Society and set up his office in the MPI-M in Hamburg, where he was a valuable advisor not only to Klaus Hasselmann but also to younger colleagues. (See his interview with von Storch and Hasselmann (2003) about the foundation of the MPI-M).

Hans Hinzpeter (1921–1999) was appointed professor of physical meteorology at the Meteorological Institute of the Universität Hamburg and director at the MPI-M. As successor of Karl Brocks, he took over the group at the then-dissolved Fraunhofer Institute for Radiometeorology and Maritime Meteorology. His focus was on physical processes in the atmosphere (see his interview by von Storch and Fraedrich, 1997).

Hans Hinzpeter was succeeded by Hartmut Graßl in 1988, both as university professor at the Meteorological Institute and director at the MPI-M. Unlike Hinzpeter, Graßl was interested in climate research. Already as a research associate with Hinzpeter, he worked on radiative transfer in turbid atmospheres and clouds, and he was one of the first to introduce the assumption of maximum entropy production into climate modelling in 1981 (Grassl 1981).

The MPI-M closely cooperated with the Universität Hamburg. Apart from the link via Hans Hinzpeter and his successors at the Meteorological Institute, two other professors were significant partners, namely Günter Fischer, professor of theoretical meteorology at the Meteorological Institute and Jürgen Sündermann, at the Institute of Oceanography (von Storch and Langenberg 2019). From oceanography, several scientists came to the MPI-M, such as Ernst Maier-Reimer or Uwe Mikolajwicz; from meteorology, Erich Roeckner and Hans von Storch.

After this initial phase, in around 1982, Hasselmann initiated the build-up of a climate modelling capacity, by coupling a conventional atmospheric general circulation model with a new, advanced ocean circulation model. A key person in this endeavour was Ernst Meier-Reimer (Heinze and Hasselmann 2019),[4] who designed and developed the *Large Scale Geostrophic model* of ocean circulation (Maier-Reimer 1993) (see Sect. 5.2). Such a system was well suited to study the internal variability in the climate system and other dynamics of the system, such as multiple equilibria and abrupt changes (see Chap. 4). In pursuit of an integrative view on the Earth's climate system, Hasselmann also initiated the design of models of the carbon cycle, starting with the carbon cycle in the ocean in the ocean (HAMOCC, Maier-Reimer and Hasselmann 1987; Maier-Reimer 1993, see Chap. 3). Later, this was combined with the coupled atmosphere-ocean model (Cubasch et al. 1992).

This activity required a high-performance computing and data-archiving system designed for the use of such models and for storing the large amounts of data produced. Consequently, Hasselmann convinced the Bundesministerium für

[4] A video-interview with Ernst Maier-Reiner, is available here: https://www.youtube.com/watch?v=SJhit5oGA1s.

Forschung und Technologie and the regional partners of the Universität Hamburg, the Alfred Wegener Institute in Bremerhaven, and the GKSS Research Centre (in 2021 renamed to Helmholtz-Zentrum Hereon) in Geesthacht, together with the MPG to found and fund the German Climate Computing Centre (DKRZ) in 1988. Hasselmann became director of the DKRZ, while Wolfgang Sell was responsible for the technical set-up and operation. In this phase, a special, but important, technical problem emerged when coupling an atmosphere and an ocean model, namely that the fluxes of momentum and energy supplied from the ocean to the atmosphere, and vice versa, would not be consistent. As a temporary solution, the *anomaly coupling*, was introduced (see Chap. 5).

In 1991, the modeling activities were further bolstered when Lennart Bengtsson, the former director of the European Centre for Medium-Range Weather Forecasts (ECMWF) in Reading, became the third director at the MPI-M. Under the leadership of Bengtsson, the modelling system at MPI-M became one of the leading models worldwide. A key role in this respect was played by Erich Roeckner. The MPI-M model, named ECHAM, was used regularly in all IPCC assessments of the knowledge of climate variability and climate change. The MPI-M was involved, from the beginning, in the IPCC process through Ulrich Cubasch.

Following the success of climate modelling in producing consistent scenarios of future anthropogenic climate change and the detection of anthropogenic climate change based on Hasselmann's concepts of separating noise from the signal (Chap. 2), it became clear that for the public and policymakers, the challenge would shift to assessing the impact of climate change on georisks (natural hazards) such as storm surges, on ecosystems, on society and the economy. At this time, the gap between the need for local and regional climate change at small spatial scales and the coarse spatial resolution provided by global climate models became obvious, and the concept of *downscaling* to bridge this gap was born (Chap. 6). Related to downscaling is the *upscaling problem*: to reconstruct complete climate fields from a network of local, sparse, and noisy information contained in climate proxy data (Chap. 7).

When Klaus Hasselmann retired in 1999, he was succeeded by Guy Brasseur, previously director of the Atmospheric Chemistry Division at the National Center for Atmospheric Research in the USA. With Guy Brasseur, the so-called *Earth system modelling*, i.e. the integration of biogeochemical processes, which had been conceived by Christian Junge and started by Ernst Maier-Reimer for the ocean and Martin Heimann for the atmosphere, was intensified. Lennart Bengtsson was succeeded by Jochem Marotzke, a theoretical physicist and physical oceanographer who had previously worked and taught at the Massachusetts Institute of Technology and the Southampton Oceanographic Centre and who currently devotes his efforts at MPI-M to investigate climate system variability and to medium-range climate prediction. At that time, the MPI-M consisted of two departments (Brasseur, Marotzke) dealing with global climate dynamics and one department (Graßl) dedicated to climate processes. A further restructuring of the MPI-M took place in 2005 after the retirement of Hartmut Graßl with the appointment of Martin Claussen, previously department head at the Potsdam Institute for Climate Impact

Research and professor of climate physics at the University of Potsdam. From then on, all three departments of the MPI-M worked on global Earth system and climate system modelling. After the departure of Brasseur and the appointment of Bjorn Stevens, previously professor of meteorology at the University of California Los Angeles, the departments *Atmosphere in the Earth System* (Stevens), *Land in the Earth System* (Claussen) and *Ocean in the Earth System* (Marotzke) have co-existed since 2008 until Claussen's retirement in 2021. Together with the MPI for Chemistry, Mainz, and the MPI for Biogeochemistry, founded in 1997 in Jena, the MPI-M was one of the core institutes of the Earth system cluster of the MPG (Wax 2020).

The cooperation between MPI-M and the Universität Hamburg was not only fostered by the joint appointment of a professor of physical meteorology in the full-time position at the Universität Hamburg and at the same time director at MPI-M, but also by appointments of Klaus Fraedrich, professor of theoretical meteorology at the Meteorological Institute, and Detlef Stammer, professor of remote sensing and assimilation at the Institute of Oceanography and later director of CEN (Centrum für Erdsystemforschung und Nachhaltigkeit/Center for Earth System Research and Sustainability), as Max Planck Fellows at MPI-M.

The MPI-M and the Universität Hamburg have always closely cooperated in large projects, starting in the 1970s with a "Sonderforschungbereich" (Collaborative Research Center) on Energy, Momentum, and Matter Transport in the Ocean-Atmosphere System. Not only partners of geophysical sciences, such as meteorology, oceanography and geophysics, were involved, but also from biogeochemistry, ocean chemistry, hydrobiology and fishery sciences. This was the beginning of increasing interdisciplinary climate and Earth system science in Hamburg, which led to the foundation of the Zentrum für Marine und Atmosphärische Wissenschaften (ZMAW, Center for Marine and Atmospheric Sciences) and which culminated in the successful application for an Excellence Cluster on Climate System Science and Predictability (CliSAP) in 2007, initiated by von Storch, Claussen, Fraedrich, Marotzke and Stammer. In CliSAP and since 2019 in the follow-up Excellence Cluster on Climate on Climate, Climatic Change and Society (CLICCS), natural sciences, social sciences and economic sciences were involved. With the excellence clusters, the KlimaCampus Hamburg emerged, comprising all Hamburg institutes and institutions active in climate research. The collaboration between natural and economic sciences was visionarily anticipated by Hasselmann through his ideas on coupled models of climate, environment, and society (Hasselmann 1991).

## References

Buek HW (1826) Hamburgs Clima und Witterung. Ein Beytrag zur medicinischen Topographie von Hamburg so wie zur Climatologie von Deutschland überhaupt. nach vieljährigen Beobachteung bearbeitet von H W Buek. August Campe, Hamburg

Cubasch U, Hasselmann K, Höck H, Maier-Reimer E, Mikolajewicz U, Santer BD, Sausen R (1992) Time-dependent greenhouse warming computations with a coupled ocean-atmosphere model. Clim Dyn 8:55–69

Grassl H (1981) The climate at maximum entropy production by meridional atmospheric and oceanic heat fluxes. Q J R Meteorol Soc 107(451):153–166

Hasselmann K (1991) How well can we predict the climate crisis? In: Siebert H (ed) Institut für Weltwirtschaft an der Universität Kiel, Conference on Environmental Scarcity: The International Dimension, J.C.B. Mohr (Paul Siebeck), Tübingen, pp 165–183

Heinze C, Hasselmann K (2019) Preface: Ernst Maier-Reimer and his way of modelling the ocean. Biogeosciences 16:751–753. https://doi.org/10.5194/bg-16-751-2019

Hildebrandsson HH, Köppen WP, von Neumayer G (1890) Wolken-Atlas. Gustav W. Seitz Nachf., Besthorn Bebr., Hamburg

Maier-Reimer E (1993) Geochemical cycles in an ocean general circulation model. Preindustrial tracer distributions. Global Biogeochem Cycles 7(3):645–677

Maier-Reimer E, Hasselmann K (1987) Transport and storage of $CO_2$ in the ocean - an inorganic ocean-circulation carbon cycle model. Clim Dyn 2(2):63–90

Rosenhagen G, Schatzmann M, Schrön A (2010) Das Klima der Metropolregion auf Grundlage meteorologischer Messungen und Beobachtungen. In: von Storch H, Claussen M (eds) Klimabericht für die Metropolregion Hamburg. Springer, Berlin, pp 19–59. https://doi.org/10.1007/978-3-642-16035-6

von Storch H, Fraedrich K (1997) Interview of Hans Hinzpeter. Eigenverlag, Max-Planck Institut für Meteorologie. https://doi.org/10.13140/RG.2.2.23236.83847, English version: Niels Bohr Library & Archives, American Institute of Physics, College Park, MD USA. www.aip.org/history-programs/niels-bohr-library/oral-histories/35176

von Storch H, Hasselmann K (2003) Interview mit Reimar Lüst. GKSS Report 2003/16, GKSS Research Center. https://doi.org/10.13140/RG.2.2.22764.97928, English version: Niels Bohr Library & Archives, American Institute of Physics, College Park, MD USA. www.aip.org/history-programs/niels-bohr-library/oral-histories/33761

von Storch H, Heimbach P (2022) Klaus Hasselmann – recipient of the Nobel Prize in Physics 2021. Oxford University Press Research Encylopedia of Climate Science. https://doi.org/10.1093/acrefore/9780190228620.013.931

von Storch H, Langenberg H (2019) Interview mit Jürgen Sündermann. HZG Report, HZG. https://doi.org/10.13140/RG.2.2.12015.48807

Wax G (2020) Wissenschaft zwischen Planung, Aufgabenverteilung und Kooperation. Zum Aufstieg der Erdsystemforschung in der Max-Planck-Gesellschaft, 1968-2000, Preprint 5 des Forschungsprogramms Geschichte der Max-Planck-Gesellschaft. Max-Planck-Gesellschaft. https://doi.org/10.17617/2.3224614

# Chapter 2
# The Stochastic Paradigm of Klaus Hasselmann

**Hans von Storch, Robert Sausen, Eduardo Zorita, Martin Heimann, and Martin Claussen**

**Abstract** In the early years of climate science, the primary challenge was to describe regional climates and then piece these together to create a global overview. In the 1970s, led by Klaus Hasselmann, the focus shifted to understanding how variability forms in the climate system. This issue was addressed by identifying a small, slow dynamical core governed by the Stochastic Climate Model (SCM), with the larger part of the system acting stochastically upon this core. Later, Principal Interaction Patterns (PIPs) and Principal Oscillation Patterns (POPs) were incorporated into this concept. The SCM provided a framework for evaluating numerical experiments with quasi-realistic climate models and distinguishing between unprovoked (internal) variability and externally forced variations—a separation that was central to the "detection and attribution" problem of climate change.

## 2.1 Introduction

This chapter centres around the broad recognition of the need and usefulness of stochastic thinking within the climate community, with a significant contribution from Klaus Hasselmann, a prominent researcher based in Hamburg. Hasselmann's groundbreaking work includes introducing the Stochastic Climate model, a concept now nearly regarded as commonplace. This model serves to explain the origin of internal variability on extended time scales. Another notable achievement by Hasselmann is the development of the "Detection and Attribution" methodology. This approach involves both the detection of signals arising from specific factors affecting the climatic state and the attribution of plausible causes responsible for the formation of such signals within a noisy environment.

von Storch (2022b) offers a comprehensive overview of Klaus Hasselmann's work, accompanied by insights from contemporary witnesses such as Peter Lemke, who provides an account of the background and significance of the Stochastic Climate model, shedding light on its role in the history of ideas and its importance in understanding climate dynamics.

 
H. von Storch et al., *Climate Science Concepts Born in Hamburg*,
https://doi.org/10.1007/978-3-031-81108-1_2

**Fig. 2.1** Document of the Nobel prize in Physics in 2021

The Achievement Award from the International Meeting on Statistical Climatology.[1] in 2007 recognized Klaus Hasselmann's significant contributions to the field, particularly for developing innovative ideas. In 2010, Hans von Storch received the same award for implementing some of these ideas. Later in 2021, the global significance of these contributions was underscored when Klaus Hasselmann was awarded the Nobel Prize in Physics[2] (Fig. 2.1). The enduring success of these concepts is evident in the continued widespread use of the textbook by von Storch and Zwiers (1999).

## 2.2 A Brief Account of the History of Ideas of "Models"

Creating an understanding of the dynamics of the climate system requires, first of all, observational evidence (or data) as well as theoretical understanding, which allows the construction of models, which describe the dynamics of the climate system as given by the theoretical understanding, constrained by the empirical evidence. The term "model" itself is used very differently in different quarters of science (e.g., Müller and von Storch 2004). In climate science, the first models were of the minimum-complexity variant, i.e., mathematically formulated considering

[1] http://imsc.pacificclimate.org/awards.shtml.

[2] See also Chap. 1, his Nobel speech on https://www.youtube.com/watch?v=UF8PBXkCtko, and a contextualization by, for instance, von Storch and Heimbach (2022).

only the most significant processes after strong simplifications. Other models, used in oceanography or meteorology before electronic computers became available in the late 1940s, were physical or mechanical analogues—which we sketch in Sect. 2.2.1.

When these electronic devices emerged, climate scientists took the opportunity to employ theoretical understanding to build numerical models—which required first to simplify the complete set of hydro- and thermodynamical equations and to discretise them so that they could be integrated on such computers (Sect. 2.2.2). With such models, named "maximum complexity models" by Müller and von Storch (2004), it became possible to integrate the equations forward in time—to forecast weather and simulate the climate system. Ever since, the power of computers has increased massively, which led to the challenge to improve the models, both by increasing the space-time resolution but also by including more and more processes and climate subsystems. This was about the state of affairs when, in 1975, the Max-Planck-Institute for Meteorology was established in Hamburg with Klaus Hasselmann as its founding director.

### *2.2.1 Modelling Without Electronic Computers*

Before electronic computers became widely available, the role of models was different to that nowadays. They were summaries of a concept of reality, featuring only the most relevant processes. They were used for conceptualisation, such as Niels Bohr's atom model (Hesse 1970), for forecasting, and for experimentation. Thus, the models could take on analytical expressions, manual computations involving a large number of humans named computers, or mechanical devices used in engineering. Famous is the vision of Lewis Fry Richardson of a "Forecast Factory", where large groups of human computers would be coordinated to do the necessary calculations to numerically integrate the equations of motions (Richardson and Lynch 2007). Also, Walter Hansen (see Sect. 5.1) in Hamburg, and likely many others, employed a staff of *human calculators*.

A simple mechanical device served as a "model" for the stability of the sea bed in the presence of bottom shear stress generated by waves and/or currents. In most cases, this shear stress is generated by turbulent water motion. When there is no turbulence, the surface will remain at rest, whereas heavy turbulence will cause sediment particles to become eroded and resuspended. To investigate the relationship between turbulence and sediment erosion, a simple experimental physical "model" is sketched in Fig. 2.2 (von Storch 2001). It is based on the *conceptual model* according to which turbulence is the key process; the stress associated with the turbulence causes the collapse of the surface adhesion reinforced by the presence of diatoms. The device involves taking a representative sediment sample from the region of interest and placing it at the base of a transparent tube that is then filled with water. A propeller is installed above the sediment, and its rotation speed is adjustable to regulate the turbulence intensity. Below a certain speed

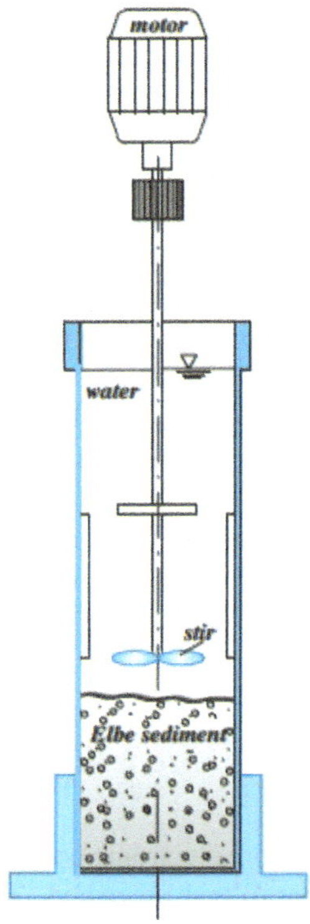


**Fig. 2.2** Mechanical device to model the effect of turbulence on sediment erosion (Schünemann and Kühl 1993)

threshold, the water column remained clear, but above this threshold, individual particles could be seen floating in the water. At the highest rotation speed used, the lower portion of the water column becomes opaque. At this point, not only the top layer of the sediment is re-suspended, but also the consolidated sediment below is starting to mobilise. Once the propeller is switched off, the suspended particles gradually settle down again.

This model does not *explain* the value of the threshold, or how deep the eroded layer is. It "only" informs about the existence of a threshold and it allows to determine these critical numbers. To constitute "understanding" we need to abstract from the concrete set-up, and design a conceptual model, a "Gedankenmodell" (mental model; Müller and von Storch 2004). The laboratory model then serves two functions, first to validate the conceptual model, and, second, to estimate unknown *parameters*.

Section 5.1 sketches an expensive and inflexible hydraulic model details to investigate tidal dynamics in the Jade estuary located in Germany. In the 1950s, ocean scientist Walter Hansen (see Sect. 5.1) recognised the potential of numerical computer models. It turned out that the much cheaper numerical models were more powerful than hydraulic models.

### 2.2.2 *Modelling with Electronic Computers*

The early development of numerical models of the *general circulation of the Atmosphere* right after World War II is described by Fleming (2011). Such models are the key building blocks of contemporary quasi-realistic climate models. When the first computers were set up, mostly for the US military, meteorologists considered using them, and decided "*to emphasize first the theory of the general circulation rather than the more applied kinematic problem of weather forecasting*". There was certainly a genuine interest in quantifying and eventually understanding the general circulation of the atmosphere, but that was not all, as Carl Gustav Rossby explained in 1946 in a letter that employing the ENIAC machine "*could be used to solve numerically the problem of the circulation of the atmosphere ... it should perhaps be possible to see to what extent various arbitrarily introduced changes in the amount and character of incoming solar radiation would affect the general circulation pattern; this would certainly be an important first step towards long-range forecasting of climatic fluctuations ...*". Scientists in those days had weather modifications in mind, or, in modern terminology, geoengineering. A decade later, Harry Wexler (Fleming 2011) stated "*When serious proposals for large-scale weather modifications are advanced, as they inevitably will be, the full resources of general-circulation knowledge and computational meteorology must be brought to bear in predicting the results as to avoid the unhappy situation of the cure being worse than the ailment.*" In any case, the motivation for developing such numerical dynamical models of atmospheric dynamics was far from only being directed towards weather forecasting.

The initial hurdles involved simplifying the equations of motion into a numerically integrable form, exemplified by the *p*rimitive equations and devising suitable discretizations. This groundwork laid the foundation for constructing models depicting the broader atmospheric and oceanic circulation, facilitating the simulation of weather patterns over prolonged periods—essentially, the study of climate. A prominent figure in this field was Syukuro Manabe, another recipient of the Nobel prize in physics for his pivotal role in developing complex climate models (Navarra 2022).

When the Max-Planck Institute of Meteorology was established, two Hamburgian professors, Jürgen Sündermann, and Günter Fischer, together with Erich Roeckner and Ernst Maier-Reimer were engaged in constructing such "quasi-realistic" models in oceanography and meteorology.

## 2.3 "Need First to Think About Natural Variability"

When Klaus Hasselmann set up his Max-Planck Institute, "*everyone expected us to immediately buy a huge computer and start calculating*" with such quasi-realistic models.[3] He did not. When "*asked whether we needed a computer*" he responded "*that I did not need a large computer, but would want one later. First, we would need to develop our research program. It was clear to me that we had to solve many fundamental issues first. ... When the institute was created, I had two goals. One was understanding the origin of the natural variability of climate. This was not understood at all, but was clearly a key issue if we wished to distinguish between natural climate variability and human made climate change. ... The other goal was developing a good ocean circulation model for climate studies.*" He achieved both goals, and in this chapter we deal with the first goal; the second is dealt with in Sect. 5.2.

At that time, information regarding variations linked to internal dynamics, as distinct from variations influenced by external factors, was limited. Understanding these internal dynamics was crucial, serving as a fundamental prerequisite for both assessing observational data and evaluating numerical experiments using climate models. Previously, model outputs were primarily juxtaposed with observational data for short durations, often within a forecasting framework. Discrepancies between long-term observations and model results were conventionally viewed as flaws in the model. Nevertheless, it became evident that such disparities could equally indicate the prevalence of internal variability, obscuring any identifiable signals introduced by external factors in the model simulations.

### 2.3.1 Detecting the Presence of Internal Variability

The possibly first note about internal variability was reported by Chervin et al. (1974), who ran a general circulation model of the global atmosphere for an extended time with cyclo-stationary conditions[4] and noticed that mean values across the same interval exhibited different states. From this observation, they concluded that statistical hypothesis tests are needed to discriminate between "noise" and "signal" when doing experiments on, say, the effect of sea surface temperature anomalies.

In the 1980s, when ocean models could be integrated with grid resolution fine enough to allow for the formation of eddies, it became clear that also global ocean dynamics forms such internal variability. Early analyses such as the one by Cox (1987) and theoretical arguments (Holland and Haidvogel 1981) led to simulation studies such as Jochum and Murtugudde (2004) and Jochum et al. (2004) who found that internal variability can explain a significant part of the observed SST variability in the tropical Pacific Ocean and the western Indian Ocean. The surface temperature (SST) and mediate sea ice extent in high northern latitudes overlaid the Atlantic

[3] This and the next quotes from von Storch (2022b).

[4] I.e., with identical lower boundary conditions and solar insulation, apart of the annual cycle.

meridional overturning circulation sea (Kushnir 1994; Mahajan et al. 2011) varied without being pegged to some external forcing.

Penduff et al. (2018a) first gave a whole picture of oceanic chaos and its impacts, structure, and predictability (OCCIPUT), based on a long (1960–2015), large ensemble (50 members) of global ocean/sea ice $1/4°$ simulations with slightly different initial conditions. Their research shows that in the presence of mesoscale turbulence, not all oceanic signals can contribute to atmospheric variability, because in several mid-latitude areas, chaotic atmospheric processes have a stronger impact than atmospheric variability on both low-frequency variability and small-scale variability; the substantial uncertainty related to chaotic internal variability should be kept in mind when interpreting modelled and observed time series.

A similar problem for regional oceans and the atmosphere had long been overseen. It had been long believed that the constraints imposed by lateral boundary conditions would inhibit the formation of internal variability, but this was eventually proven wrong by Ji and Vernekar (1997), Weisse et al. (2000), and Rinke and Dethloff (2000) for regional atmospheres, and by Penduff et al. (2018b), Tang et al. (2020), Lin et al. (2022), and Benincasa et al. (2024) for regional oceans.

### 2.3.2 *Stochastic Climate Model*

To clarify the origin of natural variability in the climate system, Hasselmann realised that there are two types of variations in the climate system, which differ in terms of their time scales (Hasselmann 1976).[5] One type of variation is associated with long time scales, varies slowly, and is carried by only a few components of the climate system; the other type goes with short-term fluctuations and has very many degrees of freedom. The slow variations are influenced by the short-term fluctuations, not in their particular details, but in their statistics—this is the core of the **Stochastic Climate Model**.

Under a number of simplifying assumptions, namely:

- a linear system,
- the slow dimension given by just one variable,
- insignificant impact of the slow dimension on the statistics of the short-term fluctuations,

we find for the slow component

$$\frac{d\psi}{dt} = \lambda\psi + \delta \tag{2.1}$$

[5] This article is quite challenging, as is often the case with "first" articles by Klaus Hasselmann when discussing a new topic. In some cases, he later wrote a revised version with remarkable clarity. When asked to do so with this topic, he declined, because it would be "too simple", and referred to the short section in the von Storch and Zwiers (1999)-textbook.

with $\delta$ representing a random variable with a constant standard deviation, which summarizes the variety of short term fluctuations.

The significance of Eq. (2.1) consists in the fact that the spectrum of $\psi$ becomes *red* if that of $\delta$ is *white*. This finding, if not theorem, specifies that in all such systems unprovoked, i.e., unforced long-term variations will emerge—red noise—and that any analysis of variations in such a system must discriminate between internal variations and possible changes caused by external drivers. This is the Detection and Attribution problem (Hasselmann 1979a, 1979b, 1993; see below).

Even if the Hasselmann (1976)-the paper had only a few hundred citations[6] it had a profound effect. It clarified that the auto-spectrum in most climatic applications (apart from the annual and diurnal cycles) is not the sum of periodic Fourier-components—but a continuous function, namely the Fourier transform of the auto-covariance function.[7] A naive person, following the book of Shaw (1936), would expect that the spectrum of climatic variations is composed of many lines. This assumption has been falsified long ago (e.g., von Storch and Zwiers 1999). Instead, an auto-spectrum should be conceptualised in this way "*An interpretive overview of the spectrum is offered in which a distinction is made between sources of variability that arise through stochastic mechanisms internal to the climatic system (atmosphere-ocean-cryosphere) and those that arise through forcing of the system from the outside.*" (Mitchell 1976), see Fig. 2.3. The Stochastic Climate model provided a theory of the origins of the stochastic part of the spectrum, particularly for the flat parts of the spectrum between the spikes and the decrease with higher frequency.

In theoretical physics, the equation is called the Ornstein-Uhlenbeck equation. It was developed when studying the Brownian motion of heavy particles colliding with many light particles. In mathematical statistics, after discretization, it goes under the name of an auto-regressive process of 1st order or an Markov process of first order. To have the system fluctuating stationarily $\lambda \in (-2, 0]$ is required.

One might argue that these assumptions are too strong, in particular, the validity of the linearization, but in a number of cases, this approximation (2.1) turned out to summarise well the behaviour of the unforced component of the system (e.g., Lemke 1977; Frankignoul and Hasselmann 1977; Lemke et al. 1980). Two of these studies dealt with sea ice anomalies (Lemke et al. 1980), and with Sea surface temperature anomalies (Frankignoul and Hasselmann 1977). Figure 2.4 demonstrated the success of describing auto-spectra estimated from data and from a fitted AR(1)-processes (2.1). The auto spectra catch the general shape of enhanced variability on long times scales, and little variance of short time scales well. In case of the sea surface temperature, not only the spectrum of (the first EOF of) sea surface temperature, but also that of (the first EOF of) air pressure, which was used

[6] As of 25. 2024, the exact number is 269.

[7] Therefore Hasselmann told his students again and again to study the textbook of Jenkins and Watts (1968).

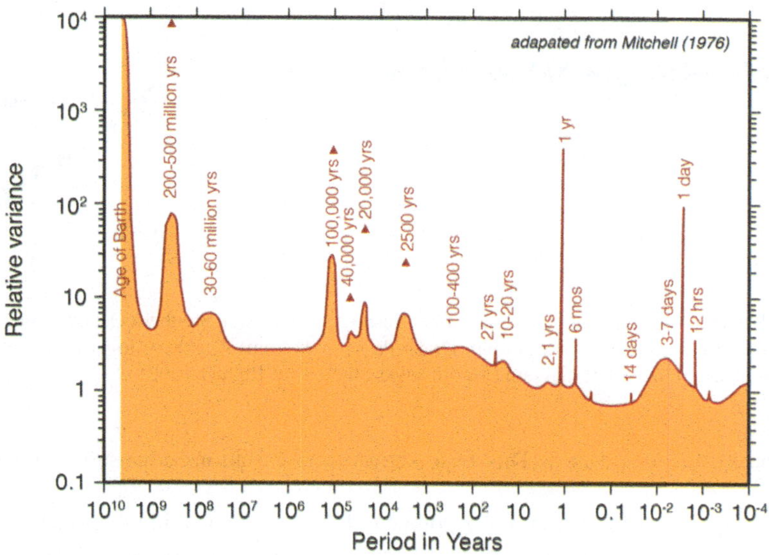


**Fig. 2.3** Conceptual proposal of the auto-spectrum of climatic variability according to Mitchell (1976). Caveat: some of the peaks would nowadays be considered as not necessarily real. 2017© Cambridge University Press, reproduced with permission

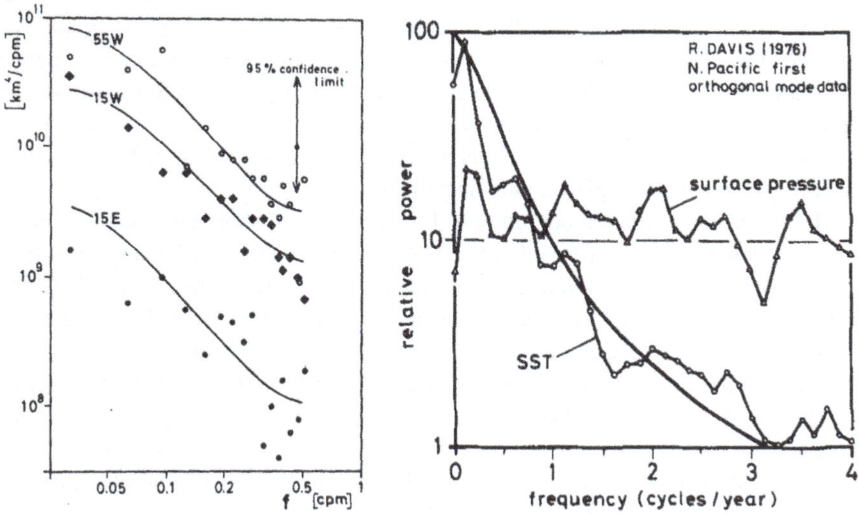


**Fig. 2.4** Estimated auto-spectra of observed data and that of a fitted autoregressive process of first order. Left: sea ice anomalies along three longitudes in the Arctic (Lemke et al. 1980) ©American Meteorological Society. Used with permission. Right: first EOF of SST in the North Pacific (Frankignoul and Hasselmann 1977)

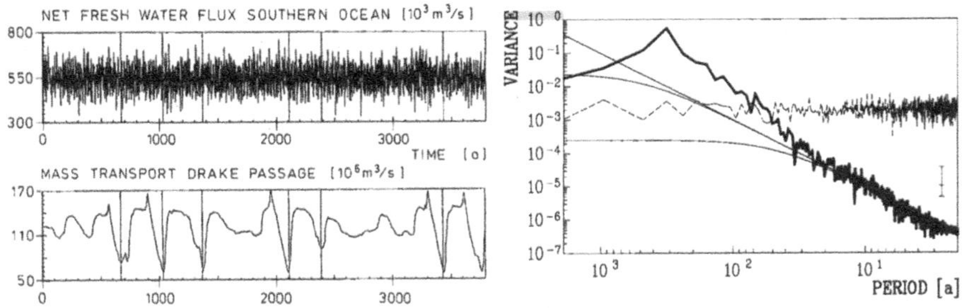


**Fig. 2.5** Left: Annual mean time series of net freshwater flux in the Southern Ocean (south of 30° S) (top) and the mass transport through Drake Passage (bottom). Right: Auto-spectra of the two curves displayed in the left diagram (Mikolajewicz and Maier-Reimer 1990)

as a proxy for heat fluxes. This EOF shows time variations consistent with white noise.[8]

A compelling illustration of the potency of the stochastic climate model emerges from a numerical experiment conducted on a comprehensive ocean model (Mikolajewicz and Maier-Reimer 1990). In this experiment, the model underwent exposure to monthly mean atmospheric surface fluxes along with the introduction of random freshwater fluxes. Although these fluxes, on average, cancelled each other out, they exhibited spatial correlation while remaining uncorrelated in time. After an extensive spin-up phase devoid of added noise, the model's equilibrium was established, and the inclusion of noise initially had minimal impact for the first few decades. However, noteworthy "oscillations" in the Southern Ocean emerged approximately 500 years later, showcasing the model's ability to generate marked climatic variations over an extended period (Fig. 2.5).

Another significant application was provided by Lemke (1977), who studied the Budyko-Sellers energy balance (see Sect. 4.2).

The stochastic climate model may well be used for all stationary, not too nonlinear systems with memory (the $\lambda\psi$-term) and forced by short-term fluctuations. If the nonlinearity becomes stronger, then the concept of "Principal Interaction Patterns" (PIPs; Hasselmann 1988) may be more suitable. The main concept behind the PIPs is to split the full state space into a small dynamically active subspace and a larger subspace whose dynamics can be parameterised as random fluctuations, as sketched in Fig. 2.6. A set of linear basis functions/patterns spanning the small dynamical subspace are named PIPs.

Even if the PIP-concept is more general and appealing, no significant practical cases concerning the climate system have emerged so far. However, when assuming a linear dynamics, at least one case has been implemented (Kwasniok 2022).

[8] At the time of the study, no long time series of reliable heat fluxes for the North Pacific were available.

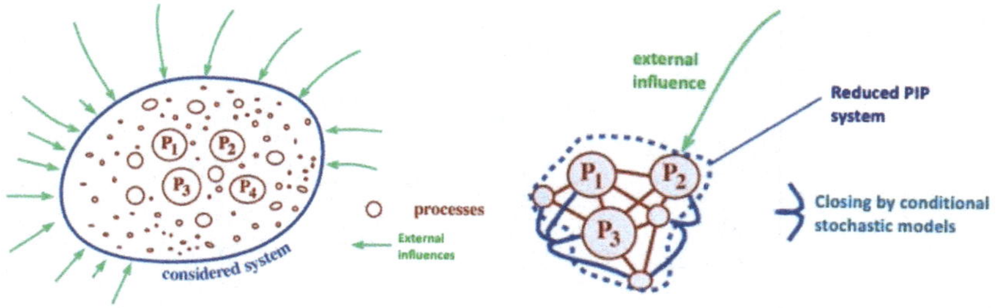


**Fig. 2.6** Sketch of the full phase space with an infinite number of processes and external forcing (left), and of the dynamically active PIP-space with a reduced number of processes and forcing, which is influenced by random parametrisations of the impacts on the large subspace (right), within which the PIP space is embedded (von Storch and Heimbach 2022)

After some implifications,[9] in particular, filtering of the full space with EOFs, the general Ansatz could be employed for multivariate state variables $\psi$ and linear dynamics. Then $\lambda$ in Eq. (2.1) becomes a matrix, whose eigenstructure determines a number of damped and oscillatory modes, eventually named "Principal Oscillation Patterns" (POPs; von Storch et al. 1988, 1995).

For this POP-case, a variety of applications have been documented in the literature. These applications are essentially a projection of the large state-space on a low-dimensional core dynamical space—which is the main concept behind the PIPs to split the full state space into a small dynamical active subspace and a larger subspace whose dynamics can be parameterised as random fluctuations. Cases refer to, for instance, ENSO, the Madden and Julian-Oscillation and its forecast, long-term oscillations in the ocean, and the life cycle of baroclinic cyclones at mid latitudes (for details and references, see von Storch et al. 1995).

In the following, we sketch the case of analysing and forecasting the Madden-and-Julian Oscillation (MJO) in some detail (von Storch and Xu 1990; von Storch and Baumhefner 1991). The MJO is a wave-1-like feature travelling eastward along the equator around the world within 30–60 days. It appears particularly clear in analyses of the velocity potential at the top of the troposphere, say 200 hPa. The POP-analysis reveals a dominant complex mode, with a wave-one structure (Fig. 2.7, top). Typically, subsequences of the quasi-periodical sequence $\cdots p_2 > p_1 > -p_2 > -p_1 > p_2 \cdots$ with a period of about 44 days emerge (von Storch and Xu 1990).

Given this tendency, the state of the POP mode is predictable for several days,—and even weeks when the mode is strong. First, the time variable coefficient, i.e., the $\gamma$ in the state $\cos(\gamma)p_1 + sin(\gamma)p_2$ is determined, and this state is predicted

[9] Disappointed by these simplifications made by Hans von Storch, Klaus Hasselmann decided to save his innovative, very general concept, by introducing a new term, namely "Principal Interaction Patterns". Eventually, both concepts were published side-by-side as Hasselmann (1988) and von Storch et al. (1988).

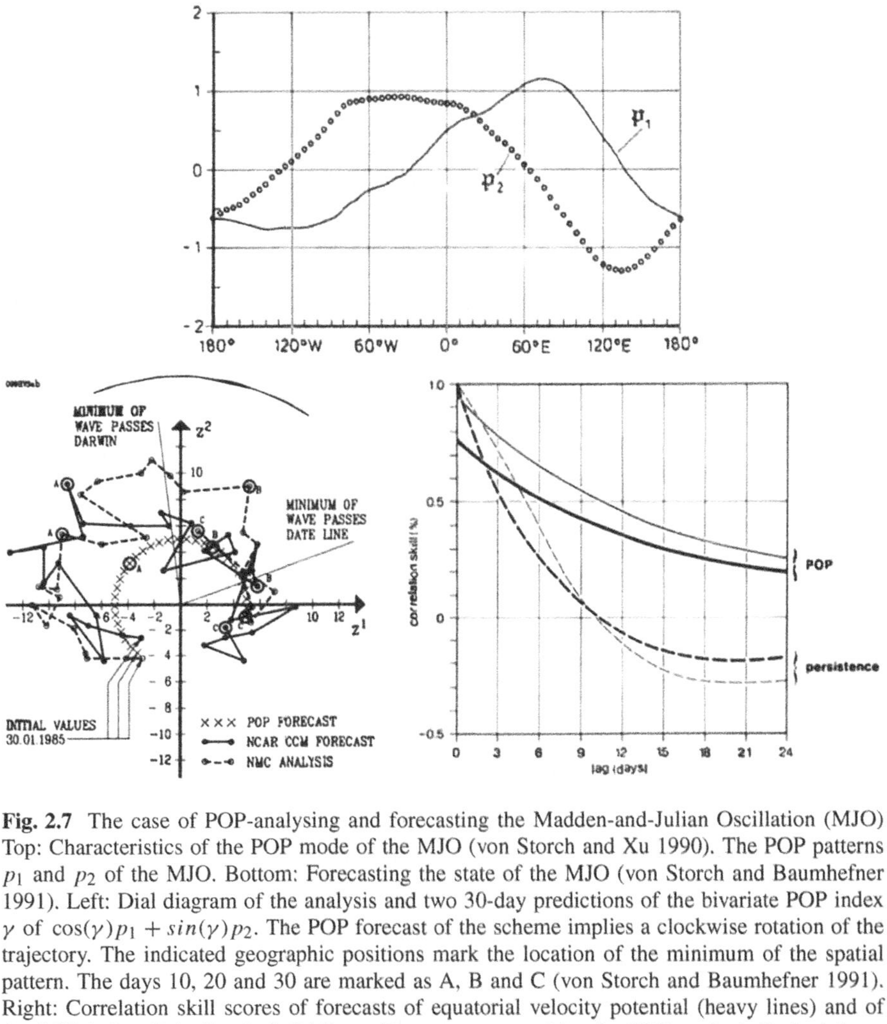


**Fig. 2.7** The case of POP-analysing and forecasting the Madden-and-Julian Oscillation (MJO) Top: Characteristics of the POP mode of the MJO (von Storch and Xu 1990). The POP patterns $p_1$ and $p_2$ of the MJO. Bottom: Forecasting the state of the MJO (von Storch and Baumhefner 1991). Left: Dial diagram of the analysis and two 30-day predictions of the bivariate POP index $\gamma$ of $\cos(\gamma)p_1 + sin(\gamma)p_2$. The POP forecast of the scheme implies a clockwise rotation of the trajectory. The indicated geographic positions mark the location of the minimum of the spatial pattern. The days 10, 20 and 30 are marked as A, B and C (von Storch and Baumhefner 1991). Right: Correlation skill scores of forecasts of equatorial velocity potential (heavy lines) and of the POP index (light lines). Solid lines: Forecasts prepared by the POP method. Dashed lines: Forecasts made by persistence

by a kind of "persistence of motion". A case with a strong initial state of the mode is shown in Fig. 2.7 (left; for further details refer to von Storch and Xu 1990). The POP forecast is given by the crosses as a pure rotation, without a change of the amplitude; a reference state of what actually happened according to the US NMC is shown by dashed lines and an experimental forecast with a full general circulation model as solid lines. The two forecasts move realistically clockwise.

An overall assessment of the skill of POP forecasts, as compared to persistence, is provided by the lower right panel of Fig. 2.7. The light curves depict the correlation skills in forecasting the POP-index, where scores below 0.5 are deemed ineffective, and a score of 1 signifies a perfect forecast. The POP method surpasses the 0.5 threshold around the 9th day, outperforming persistence, which achieves this after approximately 5 days. When extending the prediction to not only encompass the index but also include the distribution of the equatorial velocity potential anomaly at 200 hPa, the bold lines indicate that the POP forecast exhibits skillfulness for about 6 days. In contrast, persistence becomes virtually ineffectual after 3 days. Evidently, the straightforward POP forecast represents an enhancement over persistence.

**The added value of projecting high-dimensional analyses of observational data or model simulations on minimum-complexity models, such as those discussed here (Stochastic Climate model, PIPs or POPs) but also others such as Energy Balance Models is that *understanding* of the dynamics is constructed. This is a core achievement of Hasselmann's ideas, to go beyond the number crunching advanced machinery (of maximum complexity process-based models) to allow the construction of *system knowledge*.**

### 2.3.3 *Process Based Model*

Quasi-realistic models of the atmosphere or ocean dynamics, including the climate dynamics of the coupled system, are formulated using process-based dynamical equations of motion and of energy that mostly consist of conservation laws. These equations are expressed as:

$$\frac{d\psi}{dt} = \sum_i P_i(\eta) \tag{2.2}$$

where the state variables $\psi$ form a vector with very many components and the processes $P_i$ represent various physical phenomena such as advection, convection, friction, dissipation, turbulence, tides, gravity waves, and others. Often, these processes are represented by parametrizations, which provide a conditional expectation of their effects on the resolved scales.

Note that Eq. (2.2) is fundamentally different from Eq. (2.1)—the two terms on the right side do not represent processes but properties of the system, namely memory and short-term forcing, independent of which processes are responsible for generating them.

Klaus Hasselmann's concept of noise and signal underlies the implications of the emergence of internally generated variability on large scales (stochastic climate model). It clarifies the need to separate effect and noise in numerical experimentation, and in determining the presence of the impact of changing external drivers (detection and attribution of climate change).

The general strategy was to build hierarchies of models, with (2.2) as the most complex one, and Eq. (2.1) the simplest, to identify how detailed the models need to be for being able to describe certain phenomena and links. The simple models (2.1) represent understanding, while detailed simulations, forecasts and scenarios are best conducted with models of the type (2.2).

Given the noisy character of the dynamics, and the enormous number of degrees of freedom in the process-based representation (2.2), a straight-forward analysis of observed data and of simulations or numerical experiments with climate models of type (2.1) is practically impossible. In order to cast a manageable problem, one must limit the number of degrees of freedom. Or, in other words: reduce the phase space. This insight that detailed modelling must be done in the full phase space, or better, in the largest phase space available on the given computing and storing mainframe, was part of Klaus Hasselmann's thinking early on, but that the analysis needs a reduced phase space. Most of the time, he did this without making it explicit. But in 1988, he eventually formulated this principle in his PIPs and POPs-paper (Hasselmann 1988). He postulated the presence of a small signal-space described by simple equations, which governs the main aspects of the dynamics, while the rest is considered a noise-space. The statistics of the noise-space are conditioned by the state in the signal space, and the impact of the noise-space components on the signal space are realisations of these conditional statistics. The few dimensions of the signal space, given by patterns, are the Principal Interaction Patterns (PIPs). The concept is sketched in Fig. 2.3.

This approach, of splitting the full phase space into a small signal space and a large noise space permits an ordering of the dimensions of the considered system, with significant active ones in the signal space and mostly passive components in the noise space, which influence is limited to their statistics but not their state. The determination of the signal space needs a meaningful hypothesis of the key components of the system, either by using conceptual models or empirical analysis (as suggested in the PIPs and POPs paper).

In recent years, when the issue of anthropogenic climate change dominated climate science more and more, the academic interest in the question of internal variability ceased a bit. But, at least one MPI-researcher, Jin-Song von Storch, continued the quest of the role and characteristics of internal variability (von Storch 2004, 2022a) among others by demonstrating that "simulation of low-frequency climate variations and the prediction of climate change responses depend on the model representation of small-scale climate components".

From our understanding, the splitting of the phase space introduced a paradigmatic shift, made a robust analysis of climate variability and dynamical understanding possible—and, indeed, the Nobel Prize acknowledged such significant progress. In the following Sect. 2.4 the analytical practise of identifying the signal, when doing numerical experiments, with statistical hypothesis tests, is sketched, and in Sect. 2.4.1 of determining if a change in the state of the climate system may be understood as natural or must be considered at least partly anthropogenic ("detection") and, if so, which forcing emerges as most plausible ("attribution") to have lead to the change, is outlined.

## 2.4 Detecting Signals

In the early 1970s, numerical experiments could be done by a running dynamical model for extended times or by multiple,paired simulations, with one (ensemble of) simulation of a "control" and the other with a prescribed change, such as an anomalous sea surface temperature (Chervin et al. 1974). A pioneer in this direction was Bob Chervin: "*One of the key problems in analysing the results of climate experiments with general circulation models is the matter of estimating the statistical significance of a prescribed change response. This question involves separating the signal (that part of the response attributable to the prescribed change) and the noise (some measure of the inherent variability of model statistics).*" (Chervin and Schneider 1976). To address this issue, Chervin and colleagues proposed testing a null hypothesis of a mean change using a conventional Student $t$-test at all grid points.

Hasselmann noticed that, even if there would be no signal, by construction of the t-test a significant change would falsely be flagged on average at 5% of all grid points. He advocated for a more robust pattern recognition analysis, suggesting that signal and noise fields should be treated as multi-dimensional vector quantities. These fields are characterized by appropriate multi-dimensional joint probability distributions, and significance analysis should be conducted with respect to this multi-variate statistical field rather than individual gridpoint statistics. To achieve this, he recommended the use of conventional $\chi^2$ tests, later replaced by the Hotelling test (Hannoschöck and Frankignoul 1985).

Multivariate tests are effective only when the number of degrees of freedom is small; otherwise, the power of the tests diminishes significantly, making them more or less useless. Once again, Hasselmann's insight (1979a,1988) becomes relevant. He proposed that addressing this challenge requires a subdivision of the full phase space, represented by climatic variables at numerous grid points. The analysis is then conducted in a low-dimensional subspace, defined by a linear basis composed of "guess patterns," while the remaining information is disregarded as unrelated noise. These guess patterns could be constructed, for instance, from Empirical Orthogonal Functions (EOFs) or from other General Circulation Model (GCM) experiments. This approach is commonly employed when addressing the complexities of anthropogenic climate change (see below).

Hasselmann's assertion that a series of univariate tests would be inappropriate proved to be somewhat premature. This became evident when Livezey and Chen (1983) and von Storch (1982) suggested a different perspective. They proposed considering the number of univariate rejections as a random variable, with the null hypothesis positing that all "local" rejections would be false. If the tests involved are independent, a Binomial distribution can be applied in the "global" test. In cases where the tests are not independent, an attempt can be made to estimate the number of degrees of freedom in the multivariate data (Livezey and Chen 1983). This estimation can then be used as the "equivalent number of degrees of freedom" in the

Binomial distribution. This approach has become a common practice in atmospheric modelling studies.

During the early 1990s, Klaus Hasselmann successfully advocated for the establishment of a working group dedicated to the subject of "Detection and Attribution" within the US Department of Energy. This group, known as IDAG, had its program delineated by Pennell et al. (1993). A significant milestone of this initiative was the publication of the comprehensive review paper Barnett et al. (2005), summarizing the state of the art in the field.

### 2.4.1 Detection and Attribution

When examining General Circulation Model (GCM) experiments, the main hurdle lies in identifying a "significant" response, as discussed previously. Once a noteworthy effect linked to the introduced prescribed change is pinpointed, assigning it to a cause becomes straightforward—it is an outcome of the imposed alteration. However, this process often fails when dealing with the observational record. If the record includes possible signals associated with a variety of factors, such as the repercussions of transient changes in land use change and atmospheric concentrations of greenhouse gases and aerosols, attributing these signals to specific causes becomes more intricate.

The "Detection and Attribution" ansatz, as proposed by Hasselmann (1979b, 1993), involves the assumption of a small subspace wherein the significant dynamics unfold. This D&A shorthand stands for "detecting non-internally generated change" and "attribution of such change to most plausible causes".

Often, claims regarding the identification and ascription of anthropogenic climate change pivot on designating a trend as "significant." Nevertheless, this argument is frequently problematic. Technically, it is often inaccurate, as tests commonly presume serially independent data, which is not the norm for most datasets. Furthermore, from an argumentative viewpoint, the rationale is flawed since a trend can arise from natural causes, as illustrated by the Stochastic Climate model (cf. Sect. 2.3.2).

#### 2.4.1.1 Detection

The detection phase adopts the structure of a statistical hypothesis test, wherein the null hypothesis posits that the recent change originates from a distribution of changes in an undisturbed climate. Upon rejecting this hypothesis with a minimal risk of error, the alternative hypothesis is embraced. According to the alternative hypothesis, the climate responsible for the recent change is not undisturbed but influenced by an external forcing. To enhance the a-priori likelihood of successfully refuting the null hypothesis, a substantial reduction in the number of degrees of

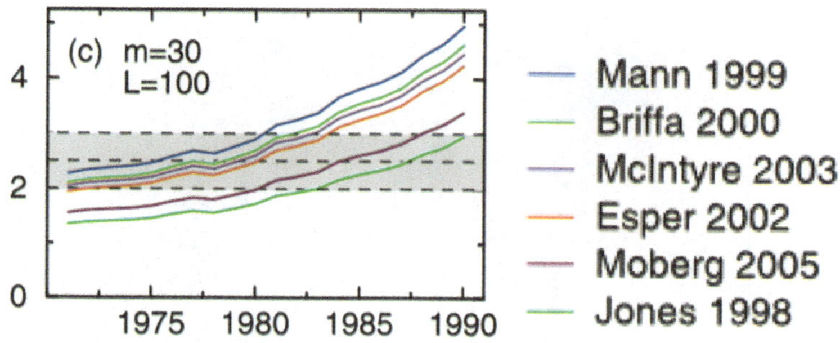


**Fig. 2.8** Smoothed and standardized instrumental Northern Hemisphere temperature developments. Shown are differences of two moving 30 year averages. The label "1990" corresponds to the more recent average, covering the period from 1976 to 2005. The temperature data are normalized by dividing them by the standard deviations of these differences, as estimated from various proxy-data reconstructions listed on the right. Thresholds of 2, 2.5, and 3 standard deviations are indicated by dashed lines (Rybski et al. 2006)

freedom is necessary. Once again, this results in a small signal space and a large noise space.

In practical terms, the signal space is defined by a limited number of anticipated responses to the presumed driver, often revolving around heightened greenhouse gas concentrations as proposed by climate models.

In the initial pursuit of assessing global anthropogenic warming, Madden and Ramanathan (1980) made an early attempt, facing challenges "possibly because the predicted warming is being delayed more than a decade by ocean thermal inertia, or because there is compensating cooling due to other factors." Another early endeavour was carried out by Karoly et al. (1994). In hindsight, analyses indicate that detection could have been feasible in the 1980s when examining changes in global latitude temperature (Rybski et al. (2006), Fig. 2.8). This assumption hinges on the reliability of proxy-based reconstructions of Northern Hemisphere temperature variations—a somewhat audacious assumption, as discussed in Sect. 7.4.1.

Klaus Hasselmann proposed augmenting the effectiveness of the statistical test by optimizing the expected signal-to-noise ratio. The idea was to identify guess patterns with maximum expected signal-to-noise ratios, a method referred to as the "optimal fingerprint" technique, first used by (Hegerl et al. 1996). However, this approach saw limited use later on, as the signal, particularly in terms of thermal variables, proved to be sufficiently strong even without this optimization.

#### 2.4.1.2 Attribution

Hasselmann's attribution phase also involves conducting a statistical hypothesis test, but the "success" is not determined by rejecting the null hypothesis; instead, it hinges on non-rejection. In other words, the recent change aligns with a climate modified by a specific combination of external forcings, as suggested by climate

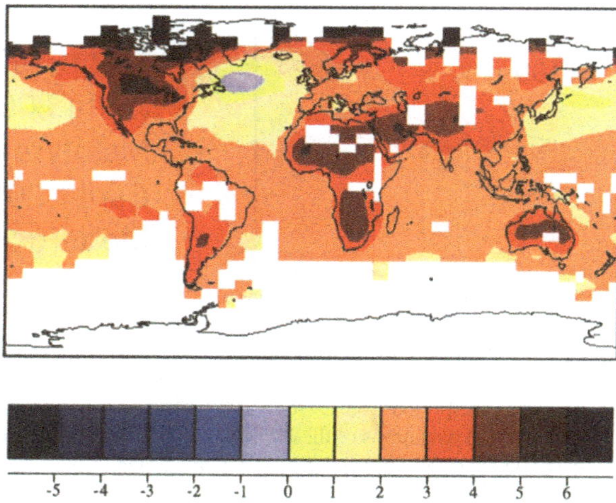


**Fig. 2.9** The normalized difference between the decades 2076–2085 and 1986–1995 derived from a simulation named "early industrial climate change" (EIN). The pattern is restricted to grid points for which reliable trends could be calculated from observations since 1949 (Hegerl et al. 1996). ©American Meteorological Society. Used with permission

models. Clearly, the evidence for attribution is less robust than the stringent detection phase. This is inevitable because it cannot be ruled out that there might be a forcing at play that science is not yet aware of.

A "detection" serves as a statistical confirmation of the null hypothesis's invalidity, indicating that a trend is inconsistent with natural or internal variability. On the other hand, an attribution is a demonstration that an observed trend is in line with a change modelled as a response to specific causes. It represents a plausibility assertion rather than a proof of validity. Therefore, a positive attribution is subject to adjustments if new mechanisms associated with the observed "detected" change are introduced or if expectations of effects from models need modification.

Hasselmann's concept saw its initial successful application documented in an MPI Report (Hegerl et al. 1994), later published in a journal (Hegerl et al. 1996). Concurrently, a group led by Ben Santer and Tom Wigley undertook a similar endeavour to detect and attribute the change in global mean temperature, utilizing spatial correlation (deviation from the spatial mean) and regression (Santer et al. 1995, 1996). Both pivotal figures, Gabriele Hegerl and Ben Santer, were working at the MPI in Hamburg during this period.

In the following, we focus on the work of Gabriele Hegerl.[10] In her first paper Hegerl et al. (1994, 1996) worked only with greenhouse-gas related change, with a pattern as suggested by a coupled simulation, named EIN ("early industrial") done

[10] See interview at http://www.hvonstorch.de/klima/Media/interviews/AS/hegerl.1102.pdf.

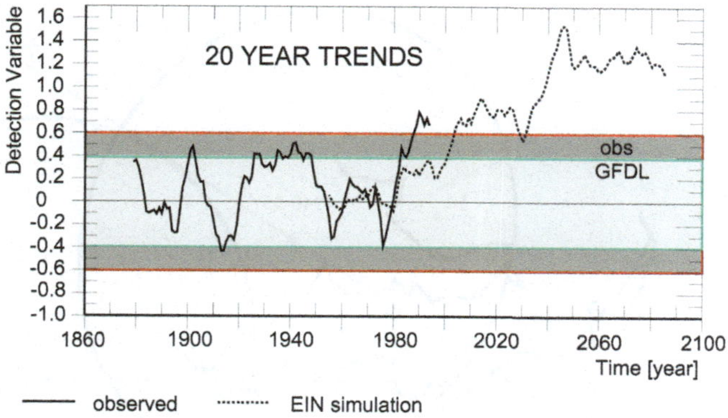


**Fig. 2.10** The evolution of the detection variable is depicted using the optimized guess-pattern fingerprint for observations (solid line) and the EIN simulation (dashed line) for trends of 20 years. The timeline corresponds to the final year of the trend. Different shadings indicate the 95% confidence intervals derived from two sets of variability data. For the one-tailed test (where the signal is known to be positive), crossing the band delineated by the green or red lines is considered evidence to reject the null hypothesis. This null hypothesis posits that the trend falls within the range of natural variations and may be rejected with a risk of 2.5% or less

with the Hamburg model named ECHAM/LSG (Fig. 2.9, Cubasch et al. 1995). The key diagram in her analysis is reprinted as Fig. 2.10.

The "detection variable" is provided as the projection on a guess pattern, representing either an expectation of the effect of enhanced greenhouse gas concentrations or an optimized pattern (Hasselmann 1993). The optimized pattern is determined as a rotation of the expected signal in a low-dimensional space spanned by the first few Empirical Orthogonal Functions (EOFs), maximizing the expected signal-to-noise ratio. To mitigate biases, the EOFs are derived from the Hamburg ECHAM/LSG model. The noise level of the detection variable is derived from either observed data or an extended simulation with a GFDL model[11] (Stouffer et al. 1994). To determine the "observed variability," the presumed signal is subtracted from the raw data.[12]

The detection variable deviates from the consistency corridor with the null hypothesis of no signal towards the end of the observed data. This outcome led to the assertion of a successful detection, contingent upon a realistic estimation of natural variability (Hegerl et al. 1996).

For comparison, the detection variable developed in the artificial world of the EIN simulation was also examined (dotted line). Also in this case, the detection variable departs from the corridor of consistency with the null hypothesis, with the deviation occurring around 2005.

[11] Both ECHAM/LSG and the GFDL employed flux correction, see also Chap. 5.

[12] For detailed information, refer to Hegerl et al. (1996); the optimized pattern is also displayed in that paper.

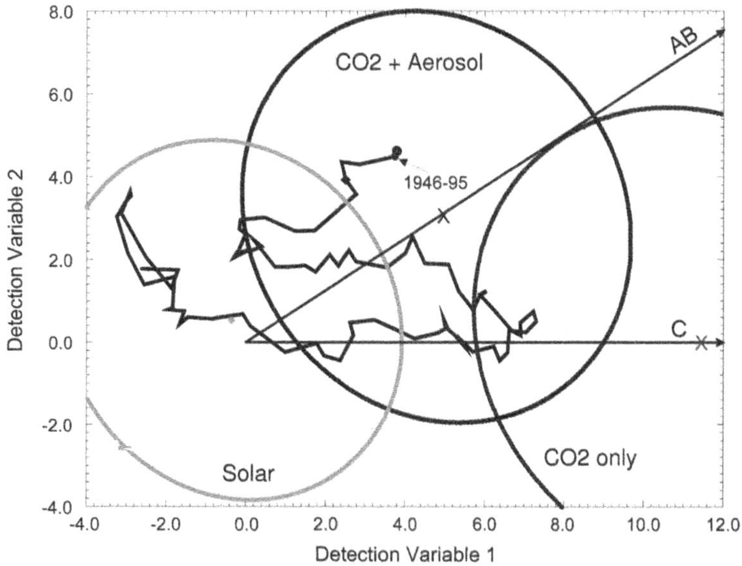


**Fig. 2.11** Development of 50 year trends in air temperature in terms of the projections on either the pure greenhouse gas signal or the combined greenhouse-aerosol signals (Hegerl et al. 1997)

In this configuration, the analysis was solely guided by the impact of elevated atmospheric greenhouse gas concentrations. Consequently, the trend in recent temperature was determined to align with the only non-natural effect in the model simulations. Given the knowledge available at that time, the attribution of the warming to the greenhouse effect was deemed valid. However, as highlighted by Hegerl et al. (1996): "*However, in order to attribute the observed warming uniquely to anthropogenic greenhouse gas forcing, more information on the climate's response to other forcing mechanisms (e.g., changes in solar radiation, volcanic, or anthropogenic sulfate aerosols) and their interaction is needed.*"

In their second seminal paper, Hegerl et al. (1997) addressed this caveat by conducting a multivariate analysis using simulations of greenhouse gas signals and combined aerosol and greenhouse gas impacts. The evolution of the two-dimensional detection variable is illustrated in Fig. 2.11. The grey ellipse encapsulates the natural variability presumed to be linked to solar and internal activity. The final trend in the observations, denoted by a "*1946–1995*", lies outside the ellipse, indicating the detection of a non-natural trend. This trend is inconsistent with a pure greenhouse gas signal but aligns with the presence of both greenhouse gas and aerosol forcing. Thus, the earlier-raised caveat was rightly acknowledged.

The D&A methodology has matured in recent years and applied to various state variables, such as temperature, rainfall, and air pressure (Barnett et al. 2005). The D&A concept played a prominent role in the IPCC assessment of the reality of human-made climate change (IPCC 1995).

### 2.4.2 *Recent Regional D&A Studies in Hamburg*

In earlier years, the analyses predominantly adopted a global perspective, possibly motivated by the political need to substantiate the reality of anthropogenic climate change and emphasise the urgency of emission reduction. This objective has since been achieved; the evidence supporting the general upward trend in temperatures because of past and ongoing emissions of greenhouse gases is widespread, and further efforts to underscore this fact are deemed unnecessary. It is now widely accepted that an explanation lacking consideration of the anthropogenic greenhouse effect is untenable.

Once the emission of greenhouse gases, as well as aerosols (more precisely, gases leading to the formation of aerosols), was clarified, public interest in D&A somewhat waned. Anthropogenic climate change has become an acknowledged fact, a household term requiring no elaborate explanation in public media. However, new interest in D&A of regional and local changes has arisen because of the need to adapt to unavoidable anthropogenic impacts. This new focus has added complexity to the D&A-challenge, because of increased noise levels and additional regional or local drivers such as land use changes (including deforestation or urbanization) or geographically scattered aerosol sources. New questions have arisen, such as the potential regional acceleration of warming due to clear air policies.

In the following, we present two examples of analyses examining the combined regional impact of heightened greenhouse gas concentrations and land use changes in Northern South America (Barkhordarian et al. 2018). Additionally, we explore the plausibility of a reduced efficiency of mitigation measures in Northern Europe, likely associated with the phase-out of aerosol emissions. The key researcher in Hamburg in this respect is Armineh Barkhordarian.

#### 2.4.2.1 Diminishing ASO Precipitation in Tropical South America

Between the years 1983 and 2012, a general decrease in precipitation was observed in tropical South America during the months of August, September, and October (ASO), as indicated by the GPCP database. This trend is evident, with the exception of localized increases over Northeast Brazil, as depicted in the left panel of Fig. 2.12 (left). The question is whether this trend is due to internal dynamics or to anthropogenic factors.

In the context of Detection and Attribution (D&A) analysis, guess patterns for the impact of increased greenhouse gas concentrations are derived by averaging the results of 29 CMIP5 simulations. These simulations involve 7 models and are conducted explicitly with historical well-mixed greenhouse gases as the forcing agent. Similarly, for the assessment of the impact of land-use change, guess patterns are derived from 10 CMIP5 simulations. This set of simulations utilizes 3 models and is conducted with land-use change as the exclusive forcing factor (Fig. 2.12, middle and left).

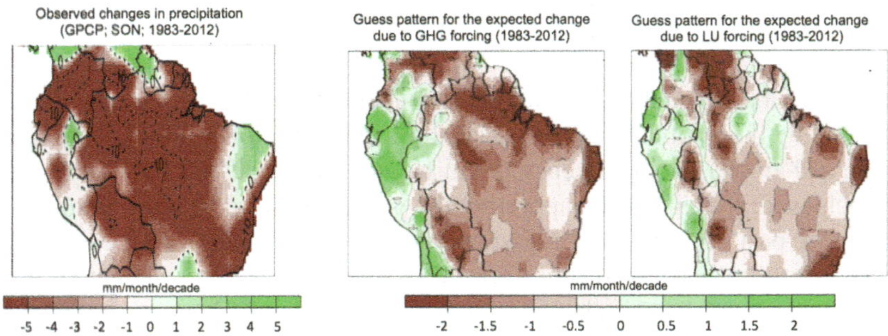


**Fig. 2.12** *Left:* Trend of changing ASO precipitation in tropical South America from 1983 to 2012 according to the GPCP database. The colour range represents the extent of the trend, ranging from green (+5 $mm^{-1}$/decade) to brown ($-5$ $mm^{-1}$/decade). *Middle and right:* Guess patterns for the expected change due to elevated greenhouse gas concentrations (GHG) and land use changes (LU). The colour range represents the extent of the trend, ranging from green (+2 $mm^{-1}$/decade) to brown ($-2$ $mm^{-1}$/decade). After Barkhordarian et al. (2018)

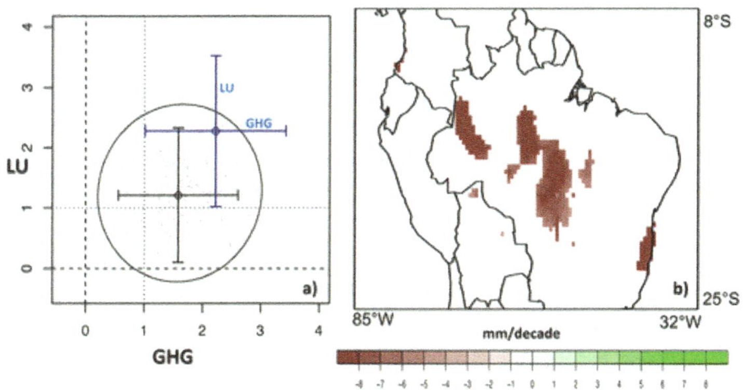


**Fig. 2.13** *Left:* The ellipse displays the joint two-dimensional 90% uncertainty interval for the greenhouse gas (GHG) and land-use change (LU) when observed data are regressed onto two guess patterns (Fig. 2.12) simultaneously during 1983–2012. The black and blue whiskers indicate the marginal bivariate and the separate univariate one-dimensional 95th %tile uncertainty intervals for the two signals, respectively. *Right:* Regions where externally forced changes are still detectable after removing the effect of GHG and LU forcing (at 5% level) (Barkhordarian et al. 2018)

In Fig. 2.13 (left), the blue whiskers represent the estimated 90% range for projecting the observed change independently onto the greenhouse gas (GHG) guess patterns and the land-use (LU) guess pattern. Notably, neither range includes zero, indicating that the observed change is inconsistent with a zero contribution from either of these two factors. Both univariate ranges encompass the value 1, suggesting the possibility of explaining the observed change with just one of the factors.

The diagram also features a black bivariate 90% ellipse illustrating the regression on both GHG and LU, along with the marginal 90% ranges as black whiskers. Once again, none of these ranges covers the point (0,0), signifying a successful detection of an unnatural factor. The ellipse notably includes the point (1,1), interpreted as consistency that both factors combined better represent the observed change than the regression of the separate factors shown in blue.

The authors posed an intriguing question: What remains if we subtract the change attributed to GHG and LU, and does it contain a detectable signal? In Fig. 2.13 (right), the red areas indicate significant changes when subtracting GHG and LU contributions univariately at all grid points, covering about 8% of the total area. This relatively small percentage of local rejections aligns with the possibility of all local rejections being false, as discussed earlier (von Storch 1982).

**This example holds significance as it reveals that a regional anthropogenic factor, specifically land-use change, is diagnosed to impact the regional climate with comparable intensity to the global warming attributed to greenhouse gases.**

#### 2.4.2.2 Underestimated Summer Warming in the Baltic Sea Region

The 1984–2013 mean surface air temperature trend in the Baltic Sea region was examined by Barkhordarian et al. (2016). The grey bars in Fig. 2.14 show the

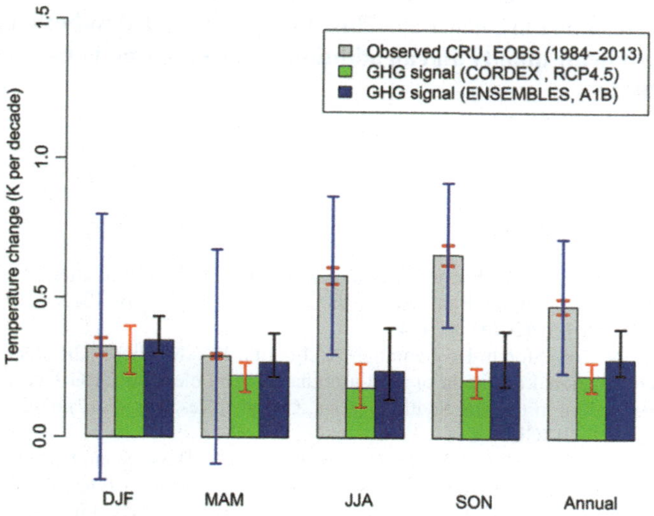


**Fig. 2.14** Observed changes in near-surface temperature within the Baltic Sea region from 1984 to 2013, represented by the grey bars juxtaposed with the mean greenhouse gas (GHG) signal estimated from nine CORDEX simulations based on RCP4.5 (depicted by green bars) and 10 ENSEMBLES projections based on SRES A1B (represented by blue bars). The brown whiskers indicate the spread of trends observed in two data sets (CRU, EOBS), while the blue whiskers denote the 95th %ile uncertainty range of observed trends derived from 1800-year historical model simulations. Additionally, the red and black whiskers illustrate the spread of trends in nine RCP4.5 and 10 A1B climate change projections. Kelvin/decade. Barkhordarian et al. (2016) CC BY-NC-ND 4.0

average of two analyses, namely CRU and EOBs. Brown whiskers mark the difference between the two, the variability give as blue whiskers is the 95% uncertainty as derived from a 1800-year historical simulation. Obviously, the two analysis are quite similar, with very short brown whiskers; however, the variability suggested by the historical simulation is wide, so the change in boreal winter (DJF) and spring (MAM) is not inconsistent with the null-hypothesis of purely natural variations, in contrast to JJA and SON, when the blue whiskers do not cover the zero.

In blue and green are the ensemble mean trends from two families of climate change simulations, with their intra-ensemble spread as read and black whiskers. The grey bar is inside the bars for DJF and MAM, but not for JJA and SON. Instead, the simulated warming is considerably smaller in summer and fall than what has been filtered from the observations.

In summary, the analysis suggests that the observed winter and spring warming, while closely aligning with model predictions, may not necessarily be of anthropogenic origin. This finding, however, does not contradict the hypothesis that elevated greenhouse gas levels contribute to this warming. On the other hand, the substantial actual warming observed in summer and fall exceeds what the models project in response to elevated greenhouse gases. While part of this warming may be attributed to such forcing, a significant portion is likely derived from other sources. One speculative explanation could be the effective reduction of aerosols in the troposphere, resulting in cleaner air. **This shift may have led to both a noticeable improvement in air quality and an additional warming trend observed between 1984 and 2013**.

## References

Barkhordarian A, von Storch H, Zorita E, Gómez-Navarro JJ (2016) An attempt to deconstruct recent climate change in the Baltic Sea basin. J Geophys Res Atmos 121(22):13,207–13,217. https://doi.org/10.1002/2015JD024648

Barkhordarian A, von Storch H, Behrangi A, Loikith PC, Mechoso CR, Detzer J (2018) Simultaneous regional detection of land-use changes and elevated GHG levels: The case of spring precipitation in tropical South America. Geophys Res Lett 45(12):6262–6271. https://doi.org/10.1029/2018GL078041

Barnett T, Zwiers F, Hengerl G, Allen M, Crowly T, Gillett N, Hasselmann K, Jones P, Santer B, Schnur R, Scott P, Taylor K, Tett S (2005) Detecting and attributing external influences on the climate system: A review of recent advances. J Clim 18(9):1291–1314. https://doi.org/10.1175/JCLI3329.1

Benincasa R, Liguori G, Pinardi N, von Storch H (2024) Internal and forced ocean variability in the Mediterranean Sea. EGUsphere 2024:1–15. https://doi.org/10.5194/egusphere-2023-3076

IPCC Second Assessment Climate Change (1995) https://archive.ipcc.ch/pdf/climate-changes-1995/ipcc-2nd-assessment/2nd-assessment-en.pdf

Chervin RM, Schneider SH (1976) On determining the statistical significance of climate experiments with general circulation models. J Atmos Sci 33(3):405 – 412. https://doi.org/10.1175/1520-0469(1976)033<0405:ODTSSO>2.0.CO;2

Chervin R, Gates W, Schneider S (1974) The effect of the time averaging on the noise level of climatological statistics generated by atmospheric general circulation models. J Atmos Sci 31:2216–2219

Cox MD (1987) An eddy-resolving numerical model of the ventilated thermocline: Time dependence. J Phys Oceanogr 17(7). https://doi.org/10.1175/1520-0485(1987)017<1044:AERNMO>2.0.CO;2

Cubasch U, Hegerl GC, Hellbach A, Höck H, Mikolajewicz U, Santer BD, Voss R (1995) A climate change simulation starting from 1935. Clim Dyn 11(2):71–84. https://doi.org/10.1007/BF00211674

Fleming J (2011) Beyond prediction to climate modeling and climate control: New perspectives from the papers of Harry Wexler, 1945-1962. In: Donner L, Schubert W, Somerville R (eds) The development of atmospheric general circulation models. Cambridge University Press

Frankignoul C, Hasselmann K (1977) Stochastic climate models, part II: Application to sea-surface temperature anomalies and thermocline variability. Tellus 29:289–305. https://doi.org/10.1111/j.2153-3490.1977.tb00740.x

Hannoschöck G, Frankignoul C (1985) Multivariate statistical analysis of a sea surface temperature anomaly experiment with the GISS general circulation model I. J Atmos Sci 42(13):1430 – 1450. https://doi.org/10.1175/1520-0469(1985)042<1430:MSAOAS>2.0.CO;2

Hasselmann K (1976) Stochastic climate models. Part I. Theory. Tellus 28:473–485

Hasselmann K (1979a) On the problem of multiple time scales in climate modeling. In: Developments in atmospheric science: man's impact on climate. Elsevier, pp 43–55

Hasselmann K (1979b) On the signal-to-noise problem in atmospheric response studies. In: Shaw B (ed) Meteorology over the tropical oceans. Royal Met. Soc., Bracknell, Berkshire, England, pp 251–259

Hasselmann K (1988) PIPs and POPs: The reduction of complex dynamical systems using principal interaction and oscillation patterns. J Geophys Res Atmos 93(D9):11015–11021. https://doi.org/10.1029/JD093iD09p11015

Hasselmann K (1993) Optimal fingerprints for the detection of time-dependent climate change. J Clim 6(10):1957–1971. https://doi.org/10.1175/1520-0442(1993)006<1957:OFFTDO>2.0.CO;2

Hegerl G, von Storch H, Hasselmann K, Santer B, Jones UC PD (1994) Detecting anthropogenic climate change with an optimal fingerprint method. MPI Report 142, Max-Planck-Institut für Meteorologie

Hegerl GC, von Storch H, Hasselmann K, Santer BD, Cubasch U, Jones PD (1996) Detecting greenhouse-gas-induced climate change with an optimal fingerprint method. J Clim 9(10):2281–2306. https://doi.org/10.1175/1520-0442(1996)009<2281:DGGICC>2.0.CO;2

Hegerl GC, Hasselmann K, Cubasch U, Mitchell JFB, Roeckner E, Voss R, Waszkewitz J (1997) Multi-fingerprint detection and attribution analysis of greenhouse gas, greenhouse gas-plus-aerosol and solar forced climate change. Clim Dyn 13(9):613–634. https://doi.org/10.1007/s003820050186

Hesse M (1970) Models and analogies in science. University of Notre Dame Press, Notre Dame

Holland WR, Haidvogel DB (1981) On the vacillation of an unstable baroclinic wave field in an eddy-resolving model of the oceanic general circulation. J Phys Oceanogr 11(4). https://doi.org/10.1175/1520-0485(1981)011<0557:OTVOAU>2.0.CO;2

Jenkins GM, Watts DG (1968) Spectral analysis and its applications. Holden-Day

Ji Y, Vernekar AD (1997) Simulation of the asian summer monsoons of 1987 and 1988 with a regional model nested in a global GCM. J Clim 10:1965–1979

Jochum M, Murtugudde R (2004) Internal variability of the tropical pacific ocean. Geophys Res Lett 31(14). https://doi.org/10.1029/2004GL020488

Jochum M, Malanotte-Rizzoli P, Busalacchi A (2004) Tropical instability waves in the atlantic ocean. Ocean Model 7(1):145–163. https://doi.org/10.1016/S1463-5003(03)00042-8

Karoly DJ, Cohen JA, Meehl GA, Mitchell JFB, Oort AH, Stouffer RJ, Wetherald RT (1994) An example of fingerprint detection of greenhouse climate change. Clim Dyn 10(1):97–105. https://doi.org/10.1007/BF00210339

Kushnir Y (1994) Interdecadal variations in north Atlantic Sea surface temperature and associated atmospheric conditions. J Clim 7(1):141–157. https://doi.org/10.1175/1520-0442(1994)007<0141:IVINAS>2.0.CO;2

Kwasniok F (2022) Linear inverse modeling of large-scale atmospheric flow using optimal mode decomposition. J Atmos Sci 79(9):2181–2204. https://doi.org/10.1175/JAS-D-21-0193.1

Lemke P (1977) Stochastic climate models, Part 3. Application to zonally averaged energy models. Tellus 29:385–392. https://doi.org/10.3402/tellusa.v29i5.11371

Lemke P, Trinkl EW, Hasselmann K (1980) Stochastic dynamic analysis of polar sea ice variability. J Phys Oceanogr 10(12):2100–2120. https://doi.org/10.1175/1520-0485(1980)010<2100:SDAOPS>2.0.CO;2

Lin L, von Storch H, Guo D, Tang S, Zheng P, Chen X (2022) The effect of tides on internal variability in the Bohai and Yellow Sea. Dyn Atmos Oceans 98:101301. https://doi.org/10.1016/j.dynatmoce.2022.101301

Livezey RE, Chen WY (1983) Statistical field significance and its determination by Monte Carlo techniques. Mon Weather Rev 111(1):46–59. https://doi.org/10.1175/1520-0493(1983)111<0046:SFSAID>2.0.CO;2

Madden RA, Ramanathan V (1980) Detecting climate change due to increasing carbon dioxide. Science 209(4458):763–768. https://doi.org/10.1126/science.209.4458.763

Mahajan S, Zhang R, Delworth TL (2011) Impact of the Atlantic meridional overturning circulation (AMOC) on Arctic surface air temperature and sea ice variability. J Clim 24(24):6573–6581. https://doi.org/10.1175/2011JCLI4002.1

Mikolajewicz U, Maier-Reimer E (1990) Internal secular variability in an ocean general circulation model. Clim Dyn 4(3):145–156. https://doi.org/10.1007/BF00209518

Mitchell JM (1976) An overview of climatic variability and its causal mechanisms. Q Res 6(4):481–493. https://doi.org/10.1016/0033-5894(76)90021-1

Müller P, von Storch H (2004) Computer modelling in atmospheric and oceanic sciences - building knowledge. Springer, Berlin

Navarra A (2022) Syukuro Manabe – recipient of the Nobel Prize in Physics 2021. Oxford University Press Research Encyclopedia of Climate Science

Penduff T, Close S, Molines JM, Barnier B, Bessières L, Maze G (2018a) Chaotic variability of ocean heat content: Climate-relevant features and observational implications. Oceanography 31. https://doi.org/10.5670/oceanog.2018.210

Penduff T, Sérazin G, Leroux S, Close S, Molines JM, Barnier B, Bessières L, Terray L, Maze G (2018b) Chaotic variability of ocean heat content: Climate-relevant features and observational implications. Oceanography. https://doi.org/10.5670/oceanog.2018.210

Pennell WT, Bamett TP, Hasselmann K, Holland W, Karl T, MacCracken GNM, Moss M, Pearman G, Rasmusson E, Santer B, Smith W, von Storch H, Switzer P, Zwiers FW (1993) The detection of anthropogenic climate change. In: Fifth Symposion on Global Change Studies. American Meteorological Society, pp 17–22

Richardson LF, Lynch P (2007) Weather prediction by numerical process, 2nd edn. Cambridge Mathematical Library, Cambridge University Press. https://doi.org/10.1017/CBO9780511618291

Rinke A, Dethloff K (2000) On the sensitivity of a regional arctic climate model to initial and boundary conditions. Clim Res 14:101–113

Rybski D, Bunde A, Havlin S, von Storch H (2006) Long-term persistence in climate and the detection problem. Geophys Res Lett 33(6). https://doi.org/10.1029/2005GL025591

Santer BD, Taylor KE, Wigley TML, Penner JE, Jones PD, Cubasch U (1995) Towards the detection and attribution of an anthropogenic effect on climate. Clim Dyn 12(2):77–100. https://doi.org/10.1007/BF00223722

Santer BD, Taylor KE, Wigley TML, Johns TC, Jones PD, Karoly DJ, Mitchell JFB, Oort AH, Penner JE, Ramaswamy V, Schwarzkopf MD, Stouffer RJ, Tett S (1996) A search for human influences on the thermal structure of the atmosphere. Nature 382(6586):39–46. https://doi.org/10.1038/382039a0

Schünemann M, Kühl H (1993) Experimental investigations of the erosional behaviour of naturally formed mud from the Elbe estuary and the adjacent Wadden Sea, Germany. In: Mehta A (ed) Coastal and estuarine studies. American Geophysical Union, pp 314–330

Shaw SN (1936) Manual of Meteorology. Volume II: Comparative Meteorology, 2nd edn. Cambridge University Press, 472 pp

Stouffer RJ, Manabe S, Vinnikov KY (1994) Model assessment of the role of natural variability in recent global warming. Nature 367(6464):634–636. https://doi.org/10.1038/367634a0

Tang S, von Storch H, Chen X (2020) Atmospherically forced regional ocean simulations of the South China Sea: Scale-dependency of the signal-to-noise ratio. J Phys Oceanogr 50:133–144. https://doi.org/10.1175/JPO-D-19-0144.1

von Storch H (1982) A remark on Chervin-Schneider's algorithm to test significance of climate experiments with GCM's. J Atmos Sci 39(1):187–189. https://doi.org/10.1175/1520-0469(1982)039<0187:AROCSA>2.0.CO;2

von Storch H (2001) Models between academia and applications. In: von Storch H, Flöser G (eds) Models in environmental research. Springer, Berlin, pp 17–33. https://doi.org/10.1007/978-3-642-59563-9_2

von Storch JS (2004) On statistical dissipation in GCM-climate. Clim Dyn 23(1):1–15. https://doi.org/10.1007/s00382-004-0404-2

von Storch JS (2022a) On equilibrium fluctuations. Tellus A Dyn Meteorol Oceanogr. https://doi.org/10.16993/tellusa.25

von Storch (ed) H (2022b) From decoding turbulence to unveiling the fingerprint of climate change: the science of Klaus Hasselmann. Springer Publisher

von Storch H, Baumhefner DP (1991) Principal oscillation pattern analysis of the tropical 30- to 60-day oscillation. part II: The prediction of equatorial velocity potential and its skill. Clim Dyn 6(1):1–12. https://doi.org/10.1007/BF00210577

von Storch H, Heimbach P (2022) Klaus Hasselmann – recipient of the Nobel Prize in Physics 2021. Oxford University Press Research Encylopedia of Climate Science. https://doi.org/10.1093/acrefore/9780190228620.013.931

von Storch H, Xu J (1990) Principal oscillation pattern analysis of the 30- to 60-day oscillation in the tropical troposphere. Clim Dyn 4(3):175–190. https://doi.org/10.1007/BF00209520

von Storch H, Zwiers FW (1999) Statistical analysis in climate research. Cambridge University Press

von Storch H, Bruns T, Fischer-Bruns I, Hasselmann K (1988) Principal oscillation pattern analysis of the 30- to 60-day oscillation in general circulation model equatorial troposphere. J Geophys Res Atmos 93:11022–11036. https://doi.org/10.1029/JD093iD09p11022

von Storch H, Bürger G, Schnur R, von Storch JS (1995) Principal oscillation patterns: A review. J Clim 8(3):377–400. https://doi.org/10.1175/1520-0442(1995)008<0377:POPAR>2.0.CO;2

Weisse R, Heyen H, von Storch H (2000) Sensitivity of a regional atmospheric model to a sea state dependent roughness and the need of ensemble calculations. Mon Weather Rev 128:3631

# Chapter 3
# The Carbon Cycle in the Climate System

**Martin Heimann, Martin Claussen, Eduardo Zorita, Robert Sausen, and Hans von Storch**

**Abstract** Envisaging the global carbon cycle as a component of the climate system was recognised already in the 1970s, but its full integration into Earth System models was achieved only after the year 2000. Landmark contributions from Hamburg in the 1980s and 1990s significantly fostered this endeavour. This chapter describes the Hamburg efforts embedded in the history of global carbon cycle research in the twentieth century: Representing the key ocean and land carbon cycle processes in a mechanistic, dynamic and geographically explicit form to be linked to quasi-realistic climate models and developing a methodology to quantify, constrain and optimise these models from global observations.

## 3.1 Introduction

The "global biogeochemical cycles" represent an interesting component within the Earth System. They describe the cycling of key elements (a.o. carbon, oxygen, nitrogen, phosphorus) through the different spheres of the Earth System by means of physical, chemical and also biological processes, which themselves are controlled by the prevailing weather, climate and environmental conditions. Some of these element cycles are reflected in the atmosphere by important greenhouse gases, e.g. carbon dioxide ($CO_2$), methane ($CH_4$) and nitrous oxide ($N_2O$), which impact the atmospheric radiation balance and therefore the prevailing climate. Hence these cycles comprise potential feedback loops in the Earth System, as is well documented by the observed tight correlation between time series of greenhouse gas concentrations and corresponding temperature proxies recorded in polar ice cores over the glacial cycles (see Fig. 3.1).

The global carbon cycle is a prominent example of a biogeochemical cycle. It is reflected in the atmosphere by carbon dioxide ($CO_2$) which is exchanged by air-sea gas exchange with dissolved inorganic carbon in the ocean and with terrestrial organic carbon in plants and soils through photosynthesis and respiration. On time scales of tens of millennia and longer, carbon is also exchanged with the lithosphere by erosion, sediment formation and volcanism.



H. von Storch et al., *Climate Science Concepts Born in Hamburg*,
https://doi.org/10.1007/978-3-031-81108-1_3

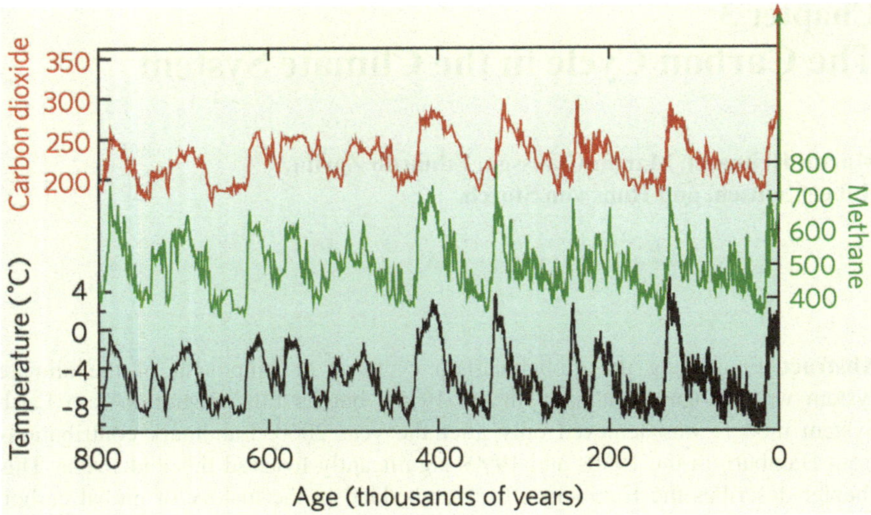


**Fig. 3.1** Correspondence between temperature and greenhouse gases ($CO_2$, Methane) over the last 800,000 years determined from the EPICA ice-core drilling project (Brook 2008, data from Lüthi et al. 2008)

Since 1750 (and perhaps also to some extent already earlier) anthropogenic activities during the industrial revolution have released large emissions of $CO_2$ into the atmosphere: primarily from the burning of fossil fuels (coal, oil and gas) and also from changes in land use (a.o. deforestation). About half of this anthropogenic $CO_2$ has accumulated in the atmosphere, raising the global $CO_2$ concentration from about 280 ppm (parts per million) in preindustrial times to almost 420 ppm today (2024), and thus contributing significantly to the recent increase in the terrestrial greenhouse effect and its associated warming climate. The other half of the emitted $CO_2$ has been taken up by the ocean and land in similar proportions.

Given this background, understanding the dynamics of the global carbon cycle in the Earth System has become an important scientific challenge. Key questions include e.g. how the anthropogenic carbon emissions are quantitatively redistributed between atmosphere, ocean and land and how this partitioning might change in a warming world. Of interest is also the strength of the coupling between the global carbon cycle and the climate system on time scales up to 100,000s of years in order to better understand the variations over the glacial cycles. And finally, from a societal point of view, understanding the direct human impacts on the global carbon cycle, primarily by the massive worldwide modifications of land use and management happening currently (a.o. deforestation, agriculture, pasture, forestry, urbanisation) constitutes a formidable task.

Models of the global carbon cycle are a key tool to address these questions. They describe the carbon cycle as a series of carbon pools connected by exchange fluxes controlled by environmental and anthropogenic factors. The mathematical formulae of these are derived from physical, chemical and biological laws or, if not available or too complicated, from empirical studies expressed as parametrisations. Evidently, a full accounting of carbon cycle dynamics within the complex Earth System requires the coupling of such carbon reservoir models into comprehensive Earth System models. This need was acknowledged already very clearly at a visionary workshop of the Global Atmospheric Research Programme (GARP) in 1974 entitled "The physical basis of climate and climate modelling", chaired by Bert Bolin and organized a.o. by Klaus Hasselmann. In a report of the meeting (GARP Joint Organizing Committee 1975) the various steps to coupling increasingly complex subsystems into a global climate model are outlined. A fundamental problem in this endeavour consists of the multiple time scales of these subsystems covering time scales too short or too large to be explicitly addressed in a single deterministic numerical model of finite size and hence need to be represented by statistically based parametrizations or specified boundary conditions, respectively. A vision of how to couple models with inherently different relevant time scales was subsequently outlined by Hasselmann (1979a). A sketch of the concept is shown in Fig. 3.2,

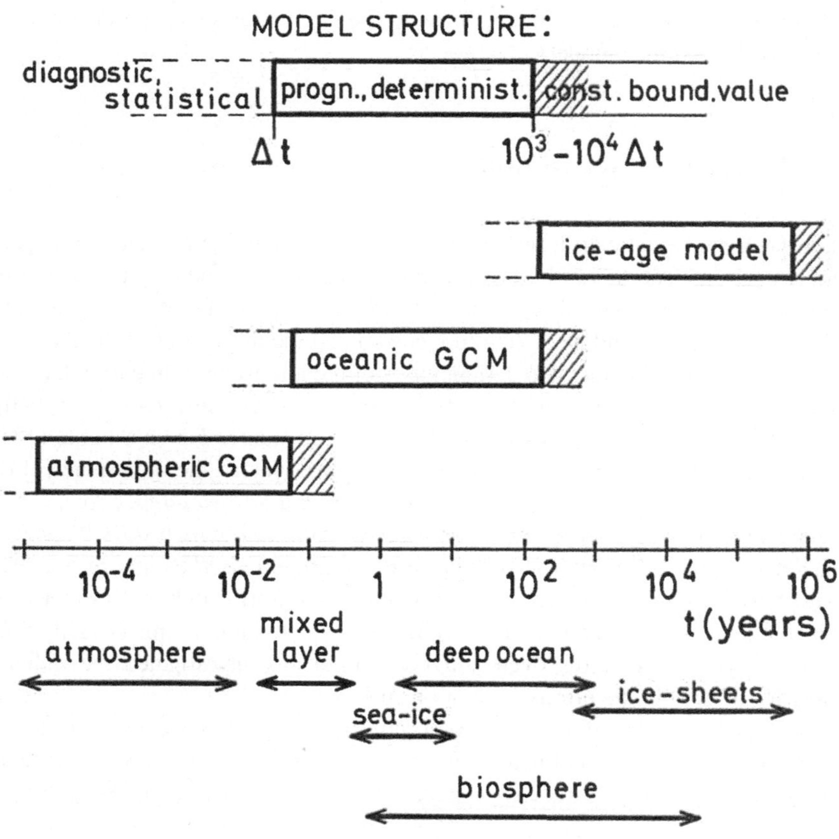


**Fig. 3.2** Natural time scales of earth system components (Hasselmann 1979a)

displaying some of the key earth system components including "biosphere" (which encompasses among others the global carbon cycle) that need to be included. Nevertheless, fully coupling the global carbon cycle into a comprehensive Earth System model took another 25 years (Cox et al. 2000).

This chapter contains a brief overview of how carbon cycle science modelling matured from using simple box models to the current generation of comprehensive global coupled carbon cycle—climate models of the Earth System. In the course of this evolution, research in Hamburg played a key, pioneering role: in establishing concepts and in developing mechanistic carbon cycle modules to be coupled to global Earth System models. After a historical note on how $CO_2$ as an important greenhouse gas got acknowledged in Earth System science, some of the early attempts to model the global carbon cycle with simple box models are described. This is followed by tracing the steps to the HAMburg Ocean Carbon Cycle model (HAMOCC), and early attempts to model similarly the land carbon cycle in a spatially explicit way. A further section deals with the atmosphere-ocean-land partitioning problem using impulse response functions. Concepts developed in part in Hamburg to use atmospheric biogeochemical observations to constrain and possibly optimise comprehensive models are described in the following section and the final section outlines the pathway to the first process-based global coupled carbon cycle—climate models.

## 3.2 Historical Background

The French mathematician Joseph Fourier (1837) was the first to recognize that the mild temperatures at the Earth's surface, compared to what would be expected from solar radiation alone, can only be explained with the help of additional heat radiating from the atmosphere and he is therefore considered the discoverer of the terrestrial greenhouse effect. The fact that certain gases of the atmosphere, in particular water vapour and $CO_2$, have the property of absorbing and emitting thermal radiation was discovered only 20 years later by the American Eunice Newton Foote (1856) with the help of a simple experiment. She clearly recognized the importance of this finding for the earth's climate: *"An atmosphere of that gas would give to our earth a high temperature; and if as some suppose, at one period of its history the air had mixed with it a larger proportion than at present, an increased temperature ... must necessarily have resulted"*. The work of Foote was unfortunately not further taken note of and was only recently rediscovered. It was the Irishman John Tyndall (1861) who three years later rediscovered and systematically investigated the radiative properties of $CO_2$ in the infrared radiation spectrum.

The quantitative relationship between atmospheric $CO_2$ concentration and air temperature near the ground was calculated for the first time by Svante Arrhenius (1896). For a doubling of the $CO_2$ concentration, he obtained a global near surface temperature increase of +5 °C to +6 °C.[1] Arrhenius was primarily concerned in his work with the search for a cause for the ice age cycles. In order to figure out how much carbon would be needed to change the atmospheric $CO_2$ concentration he turned to his colleague Arvid Gustav Högbom, a chemist, who pointed out that because of the oceans carbonate chemistry, only about 1/6 of an emitted amount of carbon would remain in the atmosphere, the rest being taken up and diluted in the world ocean. Based on this reasoning, Arrhenius considered the possible effects of anthropogenic $CO_2$ emissions on the atmospheric $CO_2$ concentration and the climate to be very small and could be detected probably only after more than 50 years.

In the 1930s it was the Briton Guy Callendar (1938) who attempted to explain the, at the time, small warming of the previous decades with an increase in the atmospheric $CO_2$ concentration. He correctly realised that on time scales up to a few centuries, it is not the entire ocean into which carbon is diluted but only the uppermost layers matter and therefore a higher fraction of any emitted fossil fuel $CO_2$ must accumulate in the atmosphere compared to Arrhenius' calculation. According to Callendars estimation, atmospheric $CO_2$ should have increased by about 18 ppm from 1900 to 1938. He supported this calculation with published chemical measurements of atmospheric air from various laboratories from the nineteenth and early twentieth centuries.[2] Callendar was certain that this increase was a result of the emissions from fossil fuel burning.

In the 1950s the topic carbon dioxide and climate attained new interest, in particular with respect to the anthropogenic perturbation. Among others, Gilbert Plass (1956) and Roger Revelle and Hans Suess (1957) provided a theoretical background of carbon exchanges between the atmosphere and the oceans, and indicated that atmospheric $CO_2$ should have increased over the industrial era, will increasingly do so due to human activities and will lead to significant climate warming by the end of the twentieth century.

The direct proof of an increase in atmospheric $CO_2$ far away from natural and anthropogenic sources and sinks was provided by Charles David Keeling (1960). With the help of an infrared spectrometer installed on the Mauna Loa volcano in Hawaii, he was able to show that in the northern hemisphere, the $CO_2$ concentration fluctuates seasonally as can be expected from photosynthesis and respiration of the northern extratropical land biosphere. From regular air samples taken over several years at the South Pole, he was also able to unequivocally show that the $CO_2$ concentration is indeed increasing worldwide. These measurements have been continued and show a steady, almost exponential increase in $CO_2$ since 1958 to almost 420 ppm today (2024), often referred to as the iconic *"Keeling curve"*, Fig. 3.3.

---

[1] This is somewhat higher than the *"very likely range"* given by the IPCC in AR6 today: +2 °C to +5 °C (AR6).

[2] Today, we know from ice core measurements that the increase of $CO_2$ in the period considered by Callendar was indeed about 14 ppm.

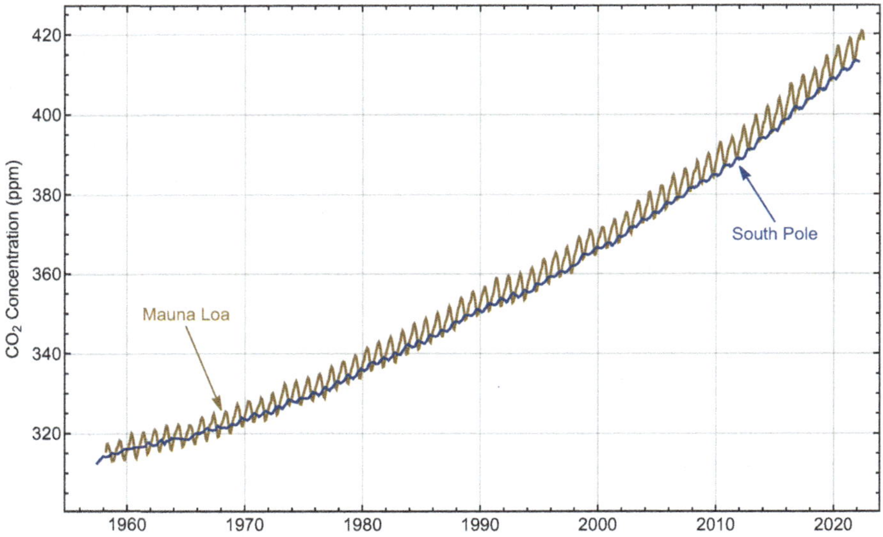


**Fig. 3.3** $CO_2$ concentration observations from direct measurements of atmospheric air at Mauna Loa and from the South Pole. In addition to the characteristic northern hemisphere annual variation and massive increasing trend, one can also see an increase in the concentration difference between the two stations. This reflects the industrial emissions located predominantly in the Northern Hemisphere: Because of the finite inter-hemispheric mixing time of the global troposphere (<1 year), this difference becomes larger as the strength of the emissions increases (Data from Scripps Institution of Oceanography; https://scrippsco2.ucsd.edu; Keeling 2022)

In the late 1960s and the following decade a network of additional observation stations were established worldwide: a.o. by the Australian CSIRO Division of Atmospheric Physics (Garratt and Pearman 1973), the Geophysical Monitoring for Climatic Change program of the National Oceanic and Atmospheric Administration (Komhyr et al. 1985), and also in 1992 by the German Umweltbundesamt (UBA) on Schauinsland). After 1980 the global network expanded further by efforts of additional institutions and became coordinated as part of the Global Atmosphere Watch (GAW) program of the World Meteorological Organization. Since 2000 it has also become possible to observe atmospheric $CO_2$ from space (column measurements): firstly from the SCIAMACHY instrument (Bovensmann et al. 1999), from the Japanese GOSAT (Yokota et al. 2004), and from the US OCO instrument (Crisp et al. 2004).

## 3.3 Early Carbon Cycle Science

Conceptually, the global carbon cycle can be separated into a "fast" component consisting of atmosphere, ocean and land surface (vegetation and soils), and a "slow" component which describes the flow of carbon between the fast component and the vast amount of carbon stored in the lithosphere through volcanism, erosion and sediment formation. For assessing the impacts of the anthropogenic emissions the "slow" component can be ignored, since the respective flow rates are two orders of magnitude smaller than the carbon exchanges between the three "fast" reservoirs. Early carbon cycle models consisted of a series of carbon reservoirs (pools, boxes) which are connected by exchange fluxes and forced by the emissions. Since carbon is a conservative quantity, only the amount of carbon (possibly in different chemical forms) is kept track of and the corresponding model equations are of the form:

$$\frac{dN_i}{dt} = \sum_j \left\{ F_{j \to i}(t) - F_{i \to j}(t) \right\} + Q_i(t) \tag{3.1}$$

where $N_i$ denotes the carbon content in reservoir $i$, $F_{j \to i}$ and $F_{i \to j}$ the carbon fluxes from reservoir $j$ to reservoir $i$ or vice versa, and $Q_i(t)$ denotes the total sources and sinks of carbon in reservoir $i$.

In this representation, the exchange fluxes are specified in highly parameterised form, depending on the carbon amount in the donor reservoir and perhaps on some additional, environmental variables, e.g. the global mean near-surface air temperature. Addressing only small perturbations on top of a steady state global carbon cycle, the models simplify to a linear ordinary differential equation system.[3] Key carbon cycle models of this kind included the models e.g. by Keeling (1973), Oeschger et al. (1975), and Bolin (1979).

If forced with exponentially rising fossil fuel emissions, these almost linear models, after proper tuning, are all able to accurately reproduce the observed atmospheric rise of the $CO_2$ concentration. In order to discriminate between different model configurations and possibly tuning some of the model parameters, simulations of carbon isotopes, in particular radiocarbon or the stable $^{13}C/^{12}C$ carbon isotope ratio were considered an essential additional model test. Obviously, these highly parameterised, tuned box-models lack a predictive capability once the carbon emissions deviate from an exponential increase and also when the environment changes. This was already clearly noted in the above mentioned GARP report of 1975. Insufficient process knowledge at that time, in particular on land carbon uptake and release, prevented the development of more comprehensive models. Therefore the need for empirical process studies in a wide range of terrestrial and marine ecosystems, as well as comprehensive global observations were strongly stressed (GARP Joint Organizing Committee 1975).

Global carbon cycle science is multidisciplinary, and combining the expertise from the different scientific disciplines is not easy. A key early German/Hamburg contribution was an international workshop held in Ratzeburg on March 21–26, 1977. It was organised by Egon T. Degens (Fig. 3.4) from the Geological-Paleontological Institute of the Universität Hamburg together with Bert Bolin,

[3] Except for the well-known carbonate chemistry of dissolved inorganic carbon in the ocean, introducing a weak non-linearity into the system.

**Fig. 3.4** Egon T. Degens. ©Photo by Stephan Kempe, Wikimedia (CC BY-SA 3.0)

Stephan Kempe and Pieter Ketner. It was attended by 66 scientists from all over the world, covering all relevant scientific domains. It provided a comprehensive review of the available scientific knowledge on the global carbon cycle, and the proceedings were published as a special report (No. 13) of the Scientific Committee of Problems of the Environment (SCOPE, Bolin et al. 1979).

The Ratzeburg workshop spurred worldwide many scientific investigations: setting up observation networks, process studies, and data compilations and fostering progress in carbon cycle modelling. In Hamburg, Egon T. Degens and his coworkers began to compile a comprehensive global data base on carbon transports by the world rivers which directly link the carbon reservoirs on land to those in the sea. Regrettably this pioneering effort was not much noticed at the time in the broader scientific community and became important only after 2010 when other groups picked up and continued the Hamburg efforts.

The Ratzeburg workshop also stressed the need to develop a unified concept for modelling carbon cycling in the Earth System. A first workshop bringing together the key carbon cycle modellers at the time was held in La Jolla in March, 1979, organised by Charles David Keeling and among others Bert Bolin. This effort resulted in a landmark publication describing the state of the art of carbon cycle modelling as a special report (No. 16) of the Scientific Committee of Problems of the Environment (SCOPE), edited by Bert Bolin (1981). This book outlined a comprehensive framework for setting up a global carbon cycle model, detailing parameter values to be used and specified even the notation to be employed for the model description. It also included key datasets (or references to them) which should be used for model evaluation. This influential book turned out to become the bible for carbon cycle modelling students in the 1980s and 1990s.

## 3.4 Complications

Until the late 1970s, global carbon cycle science was conducted primarily by a small group of geochemists, oceanographers and some atmospheric scientists. This was primarily so because carbon budget calculations at the time, i.e. the redistribution of emitted fossil fuel $CO_2$ was assumed to be primarily into the atmosphere and the ocean, and the land biosphere was considered only a "small" residual. Hence, the atmosphere-ocean partitioning was the key scientific question, which was tackled with newly available oceanographic data. In particular the ocean penetration of radioactive trace species, such as radiocarbon and tritium, which had been released as a side effect of the atomic weapon tests in the early 1960s proved to be very useful. The research field, however, changed dramatically after two major findings.

In 1978, an influential paper in Science by Woodwell and colleagues (Woodwell et al. 1978) claimed that in addition to the fossil fuel $CO_2$ there must also be huge global $CO_2$ emissions from changes in land use, predominantly from deforestation in the tropics. Indeed, their estimates ranged all the way up to 18 PgC/yr, almost double the total global emissions of today (2023)! If true, this posed a big additional question: where would then this additional $CO_2$ end up given the observed atmospheric increase and that the uptake of the ocean was considered rather well constrained? Hence a *"missing sink"* had to be found somewhere in order to balance the global carbon budget.

Evidently, a key candidate for the *"missing sink"* might be the land biosphere, i.e. the vegetation and soils not directly impacted by humans. However, this implied that land plants somehow must be stimulated to take up more carbon because of the increasing atmospheric $CO_2$ (labelled "$CO_2$-fertilization effect") or perhaps caused by a warming climate. At the time, most terrestrial ecologists were sceptical of such a mechanism since it was considered that vegetation growth is limited by available local space, climate and nutrients. Nevertheless, the quest for the *"missing sink"* widened up carbon cycle science to a host of new disciplines: terrestrial ecology, agriculture, remote sensing of vegetation etc. It also spurred a multitude of field experiments to study the effects of changing environmental conditions on the vegetation on time scales up to several decades.

A second development at the end of the 1970s/early 1980s shaking up carbon cycle science was the finding of substantially lower atmospheric $CO_2$ concentrations during the last glacial as detected in gas bubbles extracted from Greenland and Antarctic ice cores. In 1980, a Swiss (Berner et al. 1980) and a French group (Delmas et al. 1980) almost simultaneously presented the first evidence of $CO_2$ concentrations during the last glacial cold phase (about 20k years ago) of only $\sim$200 ppm, compared to Holocene/preindustrial values around $\sim$280 ppm. Over the next decades this finding was corroborated by many studies yielding a timeseries of the atmospheric $CO_2$ concentration over the last full glacial cycle; see e.g. the Antarctic Vostok record (Barnola et al. 1987). The tantalising link between greenhouse gases and climate has since been corroborated over the last 800,000 years (Fig. 3.1).

Such changes could only be caused by changing environmental conditions and thus demonstrated the need for a mechanistic explanation of carbon cycle processes on time scales from centuries to thousands of years. Obviously, this called for the development of realistic coupled climate-carbon cycle models. The finding also introduced a wide range of paleo-scientists to global carbon cycle research.

## 3.5 Ocean Carbon Cycle in a Dynamic Ocean General Circulation Model

Mechanistic modelling[4] of the ocean carbon cycle requires a physics-based ocean general circulation model in which the carbon cycle components and their chemistry are introduced as tracers. As described in Sect. 5.2, in Hamburg it was Ernst Maier-Reimer ("EMR", Fig. 3.5) who pioneered the development of the global Large-Scale Geostrophic global ocean model (LSG) which was ideally suited for long-term simulations of the ocean circulation given prescribed surface forcing fields. In parallel EMR constructed the first version of a fully fledged ocean tracer model: the *"HAMburg Ocean Carbon Cycle model"* (HAMOCC), which included a fully functional dynamic LSG model for describing the physical ocean transport and mixing. HAMOCC ultimately incorporated as tracers the various oceanic inorganic and organic carbon cycle compounds, major ocean nutrients, and it also included transient tracers, such as tritium, radiocarbon and chlorofluorocarbons.

In a landmark first publication, the LSG-HAMOCC model was used to simulate the steady state ocean inorganic carbon distribution and the inorganic carbon uptake and storage under increasing atmospheric $CO_2$ concentrations (Maier-Reimer and Hasselmann 1987). As a side result, the authors also demonstrated that the total global oceanic $CO_2$ uptake dynamics under such steady-state conditions was indeed consistent with previous box-model calculations.[5] Nevertheless, the dynamical 3-d simulation opened up a host of new opportunities to investigate the behaviour of the global carbon cycle. It allowed the confrontation of the model simulation to a vast set of available chemical and physical oceanic observations, and it permitted to study the ocean carbon cycle under changing ocean circulation and climate.

This new model attracted many global carbon cycle scientists. An influential "early adopter" was Bob Bacastow (Fig. 3.6) from the Scripps Institution of Oceanography. Staying in Hamburg for a sabbatical, he and EMR added a first version of organic carbon, nutrients and oxygen into the HAMOCC model (Bacastow and Maier-Reimer 1990, 1991; Maier-Reimer and Bacastow 1990). A later version, developed with Christoph Heinze, a student of EMR, also included a complete

[4] Christoph Heinze contributed to this section.

[5] A fact that was actually difficult for EMR to accept: he was strongly of the opinion that a physically based ocean carbon cycle model should show a different dynamical behaviour than simple global box-models.

**Fig. 3.5** Ernst Maier-Reimer (1949–2013) in his office at the MPI. ©MPI für Meteorologie

**Fig. 3.6** Bob Bacastow (1930–2007). ©SIO Photographic Laboratory Collection. SAC 44. Special Collections & Archives, UC San Diego (https://library.ucsd.edu/dc/object/bb43698099)

sediment cycle (Heinze et al. 1991). An iconic graph from an early HAMOCC simulation result is shown in Fig. 3.7; published already in Maier-Reimer and Hasselmann (1987). It displays the simulated preindustrial partial pressure of $CO_2$ ($pCO_2$) distribution in the surface ocean compared to the then-available observations. A second iconic figure stems from Christoph Heinze, showing schematically the three oceanic "carbon pumps" as implemented in HAMOCC (Fig. 3.8, Heinze et al. 1991).

The LSG-HAMOCC combination was unbeatable in the late 1980s and early 1990s as a tool for studying oceanic physical and biogeochemical dynamics and was in a completely different league in terms of throughput and usability for long-term climate studies than OGCMs based on the full "primitive" equations. This was also because in HAMOCC the biological primary production (origin of

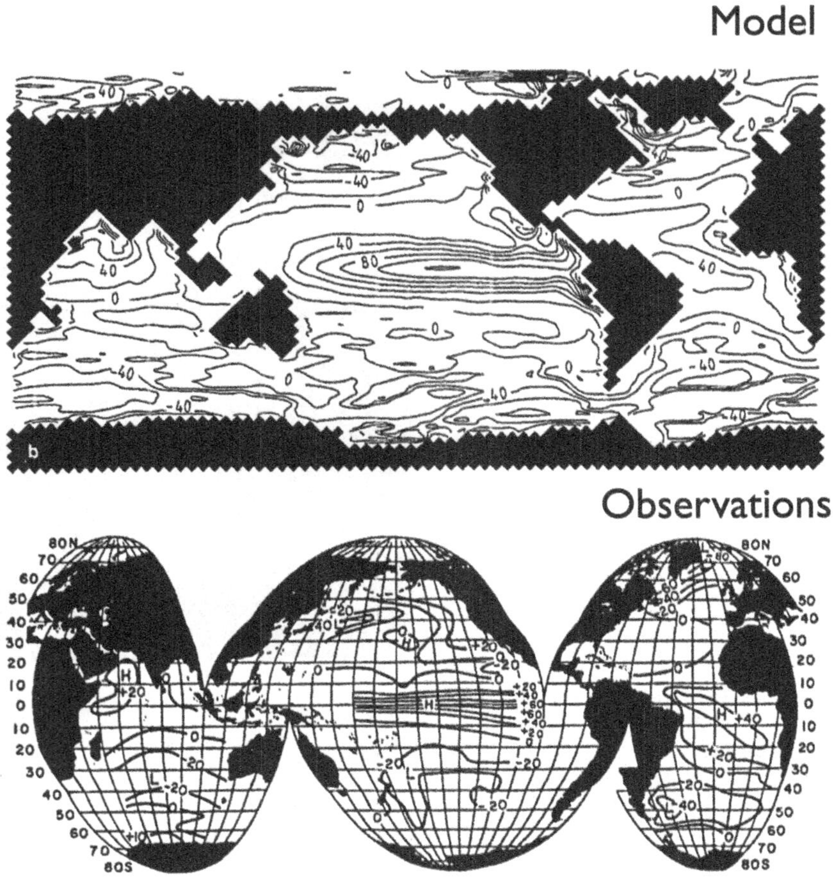


**Fig. 3.7** Preindustrial surface ocean $pCO_2$ (in $\mu$atm $\approx$ ppm, deviation from global mean) simulated by HAMOCC and as estimated from observations (Maier-Reimer and Hasselmann 1987). The contours in the upper graph are drawn with the famous plotting routine "FISLIN", also developed by EMR

the biological carbon pump) was fully prognostically determined and not through relaxation to observed nutrient fields as employed by others. One more feature of LSG-HAMOCC was absolutely essential: It featured a fabulous numerical mass conservation, needed for achieving stable global carbon cycle simulations even over 100,000 years of model integration.

Wally Broecker (Fig. 3.9, Columbia University, Lamont-Doherty Earth Observatory, LDEO) became intrigued about this ocean model after a previous LSG-HAMOCC presentation by Klaus Hasselmann at LDEO and visited the MPI for Meteorology in early 1987. Over the course of Wally's visit, some crucial improvements were made to the LSG model—also upon Wally's advice—leading Wally to state when he looked at the improved tracer simulations: "Wow, Ernst you

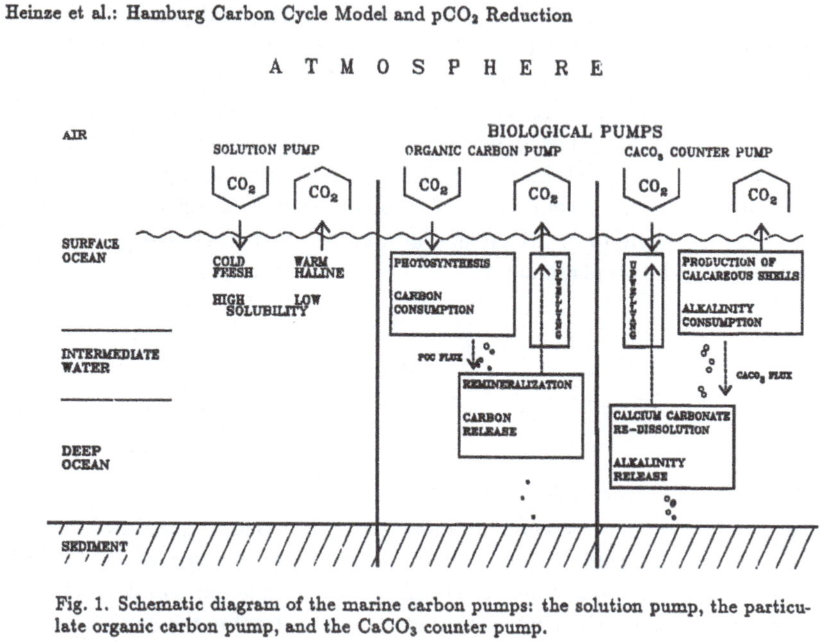


**Fig. 3.8** Schematic representation of the three ocean "carbon pumps" implemented in HAMOCC (Heinze et al. 1991)

**Fig. 3.9** Wally Broecker (1931–2019). ©Fundación BBVA Frontiers of Knowledge Awards

deserve the Nobel price for this…". The LSG and HAMOCC models were used in a series of PhD studies and provided a cornerstone for the careers of quite a number of scientists who had the privilege to use these pioneering tools when they were students.

With the LSG-HAMOCC model it was now possible to address in a mechanistic way the low glacial $CO_2$ problem: Exploring several sensitivity experiments "in qualitative agreement with marine observations" Heinze et al. (1991) showed that lowering the atmospheric $CO_2$ content in a model consistent with available glacial paleo proxy observations represents quite a challenge, and can not be achieved with a single mechanism alone. This is essentially still the view today as e.g. summarised in the 5th Assessment Report by the IPCC (see Figure 6.5 in Ciais et al. 2014).

A further pioneering application of the LSG-HAMOCC model was a first exploration of the ocean climate feedback under rising atmospheric $CO_2$ concentrations, i.e. how a changing ocean circulation in a warming climate impacts the oceanic $CO_2$ uptake and thus feeds back on the atmospheric $CO_2$ concentration (Maier-Reimer et al. 1996). Using the terminology of Friedlingstein et al. (2003, 2006) one can characterise the strength of a feedback in a model of the coupled carbon cycle—Earth System with two parameters:

$$\beta_i = \frac{\Delta N_i}{\Delta CO_2} \tag{3.2}$$

$$\gamma_i = \frac{\Delta N_i}{\Delta T} \tag{3.3}$$

where $\beta_i$ describes the change of carbon in reservoir $i$ for an atmospheric $CO_2$ increase of 1 ppm ("carbon-concentration feedback"), and $\gamma_i$ quantifies the change in carbon stock in reservoir $i$ for a global temperature increase of 1 °C ("carbon-climate feedback"). Both numbers are time scale dependent and depend to some extent on the chosen carbon emission scenario that is used in the model simulation. One can assess these feedback diagnostics from simulations performed under the different CMIPs (Climate Model Intercomparison Projects), typically evaluated from the global carbon cycle and climate changes from preindustrial until 2100. In the latest intercomparison project (CMIP6) the models exhibit values of $\beta_{\text{Ocean}} = 0.79 \pm 0.07$ PgC/ppm and $\gamma_{\text{Ocean}} = -17.2 \pm 5.0$ PgC/°C (Arora et al. 2020). Analysing the carbon inventory changes published in Maier-Reimer et al. (1996) one can estimate for LSG-HAMOCC feedback values of $\beta_{\text{HAMOCC}} \approx 1.2$ PgC/ppm and $\gamma_{\text{HAMOCC}} \approx -13.5$ PgC/°C. Thus the early HAMOCC simulation exhibits a somewhat higher ocean $CO_2$ uptake while its carbon-climate feedback is quite similar compared to the current crop of coupled ESMs.

## 3.6 Climate and $CO_2$ Driven Dynamical Terrestrial Carbon Cycle: Early Modelling Attempts

Terrestrial ecosystems are heavily shaped by the prevailing local environmental conditions. Hence, describing the land biosphere in a global earth system model requires a geographically explicit approach, which did not exist until the early 1980s. In an early pioneering study, Helmut Lieth of Osnabrück University (Fig. 3.10) had developed the "MIAMI Model" in which Net Primary Productivity (NPP), i.e. the rate of photosynthetic carbon uptake fixed as biomass in the vegetation is described as a function of local annual mean temperature and precipitation (Lieth 1975). This relationship was determined from ecological "first principles" and corroborated by

**Fig. 3.10** Helmut Lieth (1925–2015). ©Neue Osnabrücker Zeitung/Klaus Lindemann

**Fig. 3.11** Gerd Esser. ©personal communication

then available measurements from many regions of the world. Given global fields of temperature and precipitation, it became possible to predict annual NPP on a global map.

Gerd Esser (Fig. 3.11), who joined the group of Helmut Lieth in 1980, expanded this approach with climate-dependent formulations for the production and decay of litter and soil, thus closing the terrestrial cycle of carbon locally. Using this concept, Gerd Esser applied it to construct the very first, spatially explicit, "mechanistic" land carbon cycle model. In its essence, this model, termed "Osnabrück Biosphere Model" (OBM), describes in each grid cell the cycling of carbon with just four boxes, i.e. living phytomass, dead woody litter, dead herbaceous litter and organic soil carbon. Carbon flows between these reservoirs are controlled by the local climate (initially only by local annual temperature and precipitation). Additional required input information included local soil type, and the fraction of crop land

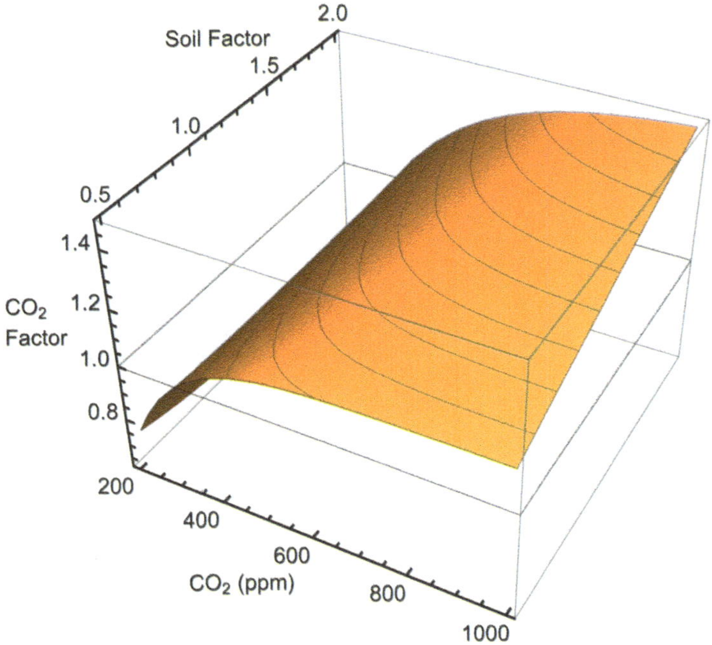


**Fig. 3.12** NPP parametrisation in the Osnabrück Biosphere Model as a function of $CO_2$ concentration and soil fertility factor (redrawn from Esser 1991)

versus natural vegetation. A key process, the dependency of NPP as a function of atmospheric $CO_2$ (i.e. the $CO_2$ fertilisation effect) is parameterised in the OBM with a simple formulation which also takes soil fertility into account as a surrogate for potential nutrient limitation of plant growth. The adopted formulation, empirically derived from field observations, is shown graphically in Fig. 3.12. It shows the expected behaviour, i.e. growth is fostered by higher $CO_2$ in a logarithmic way, and also by more fertile soils, as parameterised by a non-dimensional soil fertility factor.

The original OBM model was run on a global 2.5 by 2.5 degree latitude/longitude grid, driven by prescribed atmospheric $CO_2$, global climate fields and country-level statistics of agriculture obtained from the FAO (Esser 1984; Lieth and Esser 1985; Esser 1987). Later versions included a seasonally and interannually variable climate including humidity, additional litter fractions, vegetation fires, a more comprehensive description of agricultural techniques and would run on a global 0.5 by 0.5 degree latitude/longitude model grid. This model version got the name "High Resolution Biosphere Model" (HRBM, Esser 1995). A flowchart of the early OBM is shown in Fig. 3.13.

Obviously, the emergence of a spatially explicit land carbon model which can be driven by changes in climate and atmospheric $CO_2$ spurred a lot of interest in Hamburg and Klaus Hasselmann very early took up the opportunity and invited the

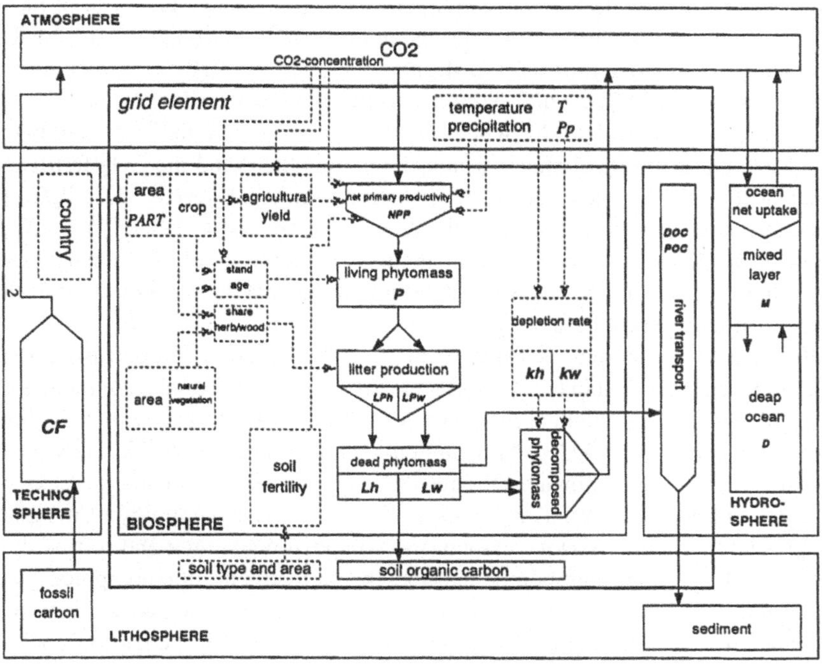


**Fig. 3.13** Schematic flowchart of the Osnabrück Biosphere Model (OBM, Esser 1987, 1991)

Osnabrück group to join the Hamburg effort of building a complete mechanistic carbon cycle description within the global climate model. With many visits of Gerd Esser in Hamburg first steps could be achieved. In particular a first version of the complete atmosphere-ocean-land carbon cycle was achieved by replacing in the OBM the original ocean box-diffusion model of Oeschger et al. (1975) with a simplified version of HAMOCC. This yielded a pioneering very first geographically explicit, off-line global carbon cycle model, i.e. an internally consistent carbon model with prescribed climate (ocean circulation, temperature, precipitation).

With the OBM-HAMOCC model, it was possible to address the global carbon budget, i.e. the partitioning of the $CO_2$ emissions between accumulation in the atmosphere, the ocean and the land. Given the highly parameterised formulation of land ecosystem processes, matching the atmospheric $CO_2$ increase, e.g. as observed at Mauna Loa, was easy given proper tuning. For this purpose, the OBM component merely confirmed previous land biosphere box-model simulations. However, the real novelty was the spatial representation predicting the key locations of the modelled land carbon sink. Since important rate constants of the model depended on local temperature and precipitation, it was also possible to investigate the climate sensitivity of the modelled terrestrial carbon uptake and decay. The climate sensitivity of the OBM model component, however, turned out to be rather small: A prescribed global linear temperature increase of +2C over the historical period until

1981 resulted in a modest enhanced terrestrial carbon uptake (Esser 1987). Indeed, based on these simulations, one can estimate that the carbon-climate feedback parameter (see Eq. (3.3)) of the OBM gives $\gamma_{\mathrm{OBM}} \approx +3$ PgC/°C. This is in stark contrast to current ESMs, which exhibit significant negative values ($\gamma_{\mathrm{Land}} = -64 \pm 70$ PgC/°C, Arora et al. 2020, which, however, were computed from simulations until 2100 under strong climate warming). On the other hand the $CO_2$-fertilisation effect of the OBM turned out to be very strong, giving $\beta_{\mathrm{OBM}} \approx 2.7$ PgC/ppm, much stronger than current state-of-the-art models ($\beta_{\mathrm{Land}} = 1.2 \pm 0.5$ PgC/ppm, Arora et al. 2020). However, current models include several additional climate-ecosystem feedback processes which substantially impact their responses and which were essentially absent in the OBM (see Sect. 3.9).

Because of the climate sensitivity of the OBM model, it was also possible to address changes in land carbon storage from the last glacial maximum to the Holocene. Esser and Lautenschlager (1994) forced the OBM with the climate from an early glacial ECHAM simulation and the corresponding low glacial atmospheric $CO_2$ and compared the resulting equilibrium land carbon stocks with the corresponding stocks under preindustrial conditions. A summary of the glacial and present (preindustrial) total carbon stocks in phytomass, litter and soils is shown in Fig. 3.14. An important component thereby to assess are the computed glacial organic carbon stocks on land areas that are subsequently inundated and lost by the rising sea level during the deglaciation. The magnitude of these carbon pools are listed in Fig. 3.14 in parentheses.

As expected, the strong $CO_2$ fertilisation effect in OBM dominates the response: It implies that the lower glacial atmospheric $CO_2$ results in substantially lower productivity and, therefore, smaller equilibrium total terrestrial carbon stocks despite

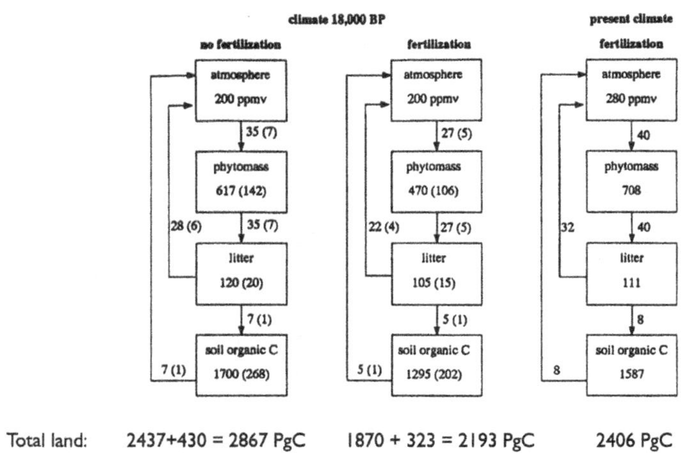


**Fig. 3.14** Terrestrial carbon stocks under glacial and preindustrial climate conditions (Esser and Lautenschlager 1994). Poolsizes in PgC, numbers in parentheses indicate the additional carbon amount on land areas during glacial conditions, which are subsequently flooded during the transition to the Holocene

the additional pools on land areas inundated during the subsequent deglaciation. Indeed, the fertilisation scenario implies a net increase of terrestrial carbon during the deglaciation of about 213 PgC, which is consistent with ice-core and sediment $^{13}C$ stable isotope data (Esser and Lautenschlager 1994). It is also consistent with current understanding (Müller and Joos 2020), even though the OBM model lacked key ecosystem processes, in particular shifts in land biome distributions which are relevant during this climate transition.

## 3.7 Synthesis of the Global Carbon Cycle in the Climate System

The link between carbon emissions -> global carbon cycle -> atmospheric $CO_2$ -> climate response contains one significant simplification: Because $CO_2$ is a long-lived atmospheric trace gas, its spatial variations are very small and hence to a very good approximation only changes of the globally averaged concentration are relevant for calculating the climate forcing and ensuing climate change given prescribed $CO_2$ emissions. As long as climate-carbon cycle feedbacks are ignored, atmospheric $CO_2$ can be seen as a single-valued function of the sum of the emissions modified by the direct response of the carbon cycle, i.e. the partitioning of the emissions into the atmosphere, ocean and land biosphere. Since this partitioning depends on the history of the atmospheric $CO_2$, and as long as the carbon cycle is assumed to be linear, the $CO_2$ concentration $C(t)$ at time $t$ can be described mathematically with a convolution integral:

$$C(t) = \int_{-\infty}^{t} h(\tau)Q(t-\tau)d\tau \tag{3.4}$$

where $Q(t-\tau)$ describes the global emissions at time $t-\tau$, i.e. $\tau$ years previous to time $t$, and $h(\tau)$ is the fraction of an initial unit emission remaining in the atmosphere at time $\tau$. $h(\tau)$ characterises the response of the global carbon cycle to emissions and can easily be computed with a global carbon cycle model. Figure 3.15 shows the impulse response functions computed with the original HAMOCC model and shows for comparison also the corresponding impulse response function of the early box-diffusion model (Maier-Reimer and Hasselmann 1987; Oeschger et al. 1975).

This representation has been successfully used by the carbon cycle research community in the past (Oeschger and Heimann 1983; Enting et al. 1994). If $h(\tau)$ of a complex global carbon cycle model is known, Eq. (3.4) provides an opportunity to represent in a very efficient way the carbon cycle dynamics to any emission history as e.g. needed in highly aggregated integrated assessment models. Furthermore, the representation (3.4) is fundamental for quantifying the global warming potential (GWP, Shine et al. 1990) where the impulse response function of $CO_2$ is used as

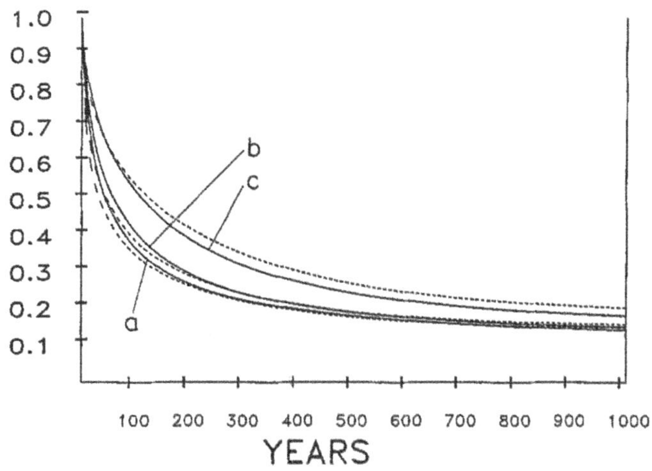


**Fig. 3.15** Fraction of carbon remaining in the atmosphere after an initial pulse emission. The solid lines correspond to different sizes of the initial pulse emissions: (a) 0.25x, (b) 2x, (c) 4x preindustrial atmospheric carbon content (Maier-Reimer and Hasselmann 1987). Dashed lines correspond to the response of the box-diffusion carbon cycle model of Oeschger et al. (1975)

reference. In most of the IPCC assessments the impulse response function of the "Bern model" was used for this purpose (Schimel et al. 1996; Joos et al. 2013).

Equation (3.4) assumes a linear carbon cycle model, hence non-linearities in the dynamics complicate the impulse response function representation. Fortunately, at least over the last six decades, the global carbon cycle has responded to the anthropogenic perturbation in a quite stable manner, as seen in the approximate constancy of the airborne fraction[6] (e.g. Bennett et al. 2024). And weak non-linearities, such as the ocean carbonate chemistry, can be simply addressed by computing $h(\tau)$ for different sizes of the pulse inputs, as also shown in Fig. 3.15. A different approach consists of specifying a base scenario and applying the impulse response formulation only for small deviations around this base case (Enting et al. 1994). It is also possible to represent only a subsystem of the carbon cycle by Eq. (3.4) and deal with the non-linearity in an explicit way (Joos et al. 1996; Hooss et al. 2001). An extension of the impulse response function representation framework has even been used to compare comprehensive ESMs, including climate-carbon cycle feedbacks (Joos et al. 2013).

[6] Commonly defined as the annual atmospheric $CO_2$ increase divided by the annual total emissions.

## 3.8 Inferring Sources from Atmospheric Observations

Spatial-temporal variations of the atmospheric $CO_2$ concentration reflect location and magnitudes of the various carbon sources and sinks at the surface of the Earth. Hence these can potentially be constrained and quantified given atmospheric observations from a global network using a model of atmospheric transport and mixing. This in turn offers the opportunity to evaluate and potentially optimise spatially explicit process-based global carbon cycle models.

Early attempts assumed that atmospheric observations of $CO_2$ reflect integrals of the sources represented only with a few meridional spherical harmonics (Bolin and Keeling 1963; Junge and Czeplak 1968). Lacking adequate meteorological data, in these studies also the magnitude of meridional atmospheric transport had to be estimated from the $CO_2$ observations. Initially with winds from a climate model (Fung et al. 1983), and subsequently with the advent of global atmospheric data assimilation products it became possible to calculate much more realistically the global time dependent three-dimensional transport and mixing (Heimann et al. 1986; Heimann and Keeling 1989).

First estimates based on the available observations in the 1980s indicated a large carbon sink in the northern hemisphere compensating in part the predominant fossil $CO_2$ emissions from the same region. This sink was postulated to be located on land (Tans et al. 1990), or also partly in the Atlantic ocean (Keeling et al. 1989). These two studies spurred a host of investigations in the 1990s to better locate, quantify and understand the nature of this "northern hemisphere sink". It also demonstrated the power of the "top-down" atmospheric inversion approach and spurred the development of a multitude of inversion systems, and ultimately also a model intercomparison project (TransCom, Law et al. 1996; Denning et al. 1999). In these studies, the globe is split into a few large-scale regions whose emissions are scaled such that the summed simulated atmospheric concentration pattern matches the observations. Since the atmospheric transport of $CO_2$ is linear, the procedure corresponds mathematically to a multiple linear regression.

Using only 3–4 global regions, Fan et al. (1998) concluded that the "northern hemisphere sink" averaged over 1988–1992 purportedly must be located primarily on the North American continent, while Eurasia exhibited only a very small carbon sink. It turned out, however, that this finding was essentially caused by the particular choice of a very small number of large-scale adjustable source regions. Indeed, with a newly developed higher resolution inversion system using an adjoint method (Giering and Kaminski 1998; Kaminski et al. 1999a,b) developed in Hamburg, Kaminski and Heimann (2001) showed that with suitable "a priori" conditions a huge hypothetical sink of 2 PgC/yr located in central Europe can be constructed, which is also fully consistent with the $CO_2$ observations from the global network (dubbed the "Empire strikes back solution", see Fig. 3.16).

This finding illustrated the under-determination of the inversion problem and the need for a careful assessment of any implicit or explicit "a priori" information underlying any inversion result. Using large source regions reduces the number

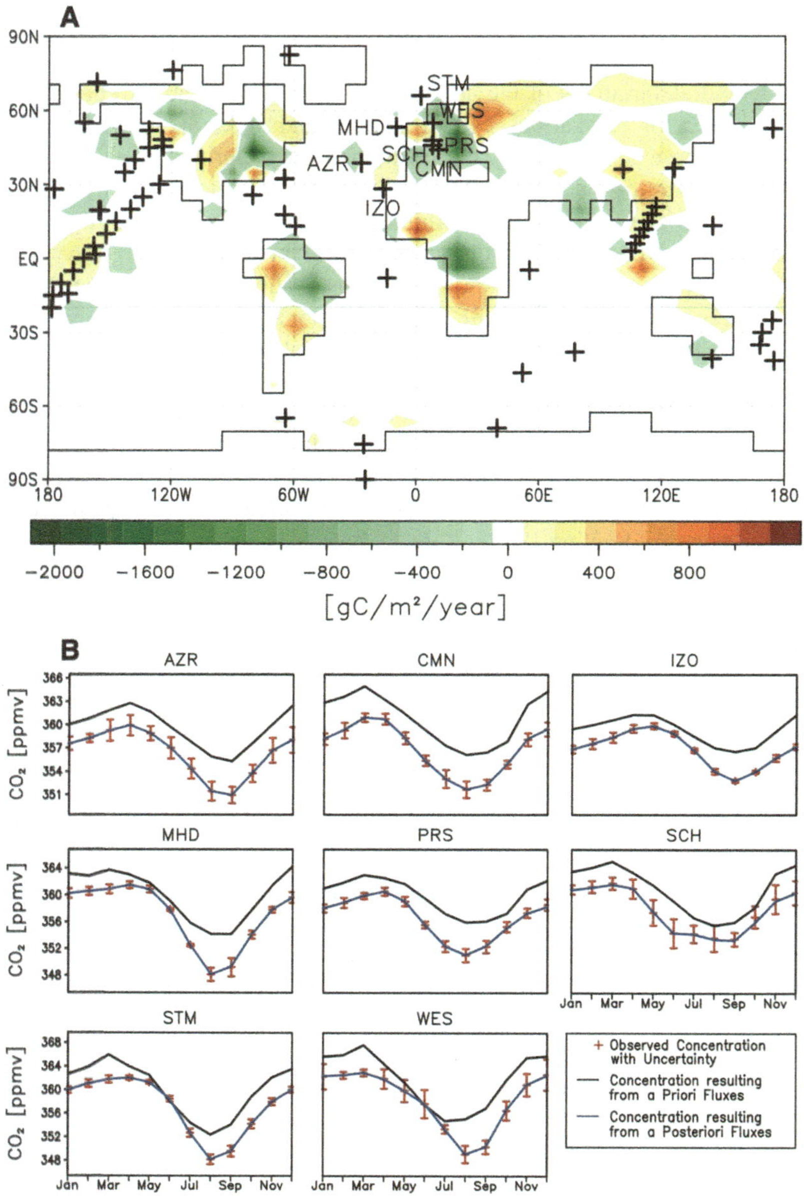


**Fig. 3.16** (**a**) Hypothetical source-sink field which exhibits a 2 PgC/yr $CO_2$ sink in central Europe, which is consistent with the atmospheric observations from the global network as shown for some stations in panel (**b**) (the "Empire strikes back solution", Kaminski and Heimann 2001)

of unknowns and stabilises the inversion, but on the other hand introduces an "aggregation error" (Kaminski et al. 2001) because of the assumed but effectively unknown smaller-scale variability structure of the sources within the regions. Ignoring this error may introduce large biases in coarse-scale inversion setups. Using a high-resolution, potentially grid-size, representation of the source fields introduces more unknowns but permits the explicit specification of the "a priori" information in order to regularise the solution (e.g. by defining spatial and/or temporal correlations between the different source elements).

Despite these difficulties, the "top-down" atmospheric inversion approach has become a powerful approach for estimating global, regional and local sources and sinks of biogeochemical species. E.g. the "Jena CarboScope" provides time-varying global maps of surface-atmosphere $CO_2$ fluxes which are routinely updated using the latest observations from the global monitoring network (Roedenbeck et al. 2018a,b, 2020). It has also become a standard method to evaluate global Earth System models or model components (e.g. Heimann et al. 1998). The method has been extended to include multiple types of observations, such as isotopic composition or other related species in addition to the $CO_2$ concentration e.g. atmospheric oxygen (Stephens et al. 1998). Furthermore, the "top-down" method is indispensable for inferring surface-atmosphere fluxes from column measurements obtained from sensors on satellites (Rayner and O'Brien 2001).

An alternative approach to address the under-determination of the inversion problem consists in solving not for the surface-atmosphere fluxes, but represents these by a surface model and uses then the atmospheric inversion system to optimise a finite number of surface model parameters (Rayner 2001). This approach has lead to the development of Carbon Cycle Data Assimilation systems (CCDAS, Rayner et al. 2005). While early studies optimised only simple surface models with very few parameters, this approach has since been extended to optimise full process-based land-surface modules of ESMs (e.g. Schuermann et al. 2016).

## 3.9 The Road to Global Coupled Carbon Cycle: Climate Models

As described above, pioneering first steps toward a geographically explicit, climate sensitive, process-based representation of the ocean and the land carbon cycle had been achieved in Hamburg and Osnabrück already in the 1980s. Nevertheless, in particular the land carbon description was not ready for inclusion as a comprehensive land carbon cycle component within a full-fledged ESM and thus fully coupled carbon cycle—climate model system simulations were not yet possible. The very simple parametrisation of photosynthesis ignored fundamental physiological knowledge, which became available in the 1980s (Farquhar et al. 1980), and the model lacked many critical processes which determine the dynamics of terrestrial ecosystems. In particular plant ecosystems are highly dependent on water, therefore any realistic physically-based representation requires an explicit description of the very fast soil-canopy-atmosphere exchanges of water and energy. On longer time scales changing environmental conditions can lead to shifts in ecosystem distributions by plant seed dispersal. Furthermore, disturbances, such as fires or insect mediated diseases can seriously impact local carbon storage in land ecosystems. Adding these missing processes into the land surface module of an ESM required to integrate essentially three modelling communities, which until the early 1990s, had worked essentially separate.

Land surface parametrisations in atmospheric general circulation models have gone through several generations of complexity from the 1960s to the 1990s in order to improve the description of surface-atmosphere fluxes of radiation, heat, water vapour and momentum (Sellers et al. 1997). In a pioneering milestone the Simple Biosphere Model (Sellers et al. 1996a,b) also included the exchange of $CO_2$ using the photosynthesis model of Farquhar et al. (1980) in order to describe realistically the plant control of the flow of water vapour and $CO_2$ through the stomates of leaves. Central objective of this land surface parametrisation, however, was a better representation of local biophysical/biophysiological feedbacks occurring on fast time scales. Hence these models did not include explicit vegetation and soil carbon pools, nor their dynamics under a changing climate. A similar "Biosphere Energy-Transfer Hydrology scheme" (BETHY) was also developed in Hamburg (Knorr and Heimann 2001a,b).

Another community at that time was concerned with assessing so-called bio-geographical feedbacks, i.e. the effects of changing surface properties caused by a changing climate driven vegetation. In order to model vegetation shifts realistically, competition between different plant species need to be taken into account. A major advancement was the concept of "plant functional types" (PFT), i.e. the reduction to a finite set of generic plant species with prescribed environmental requirements (climate, soils). A pioneering model of this kind was the BIOME 1.0 model (Prentice et al. 1992, see also Sect. 4.4). These biogeographical models were diagnostic, i.e. essentially predicting the global vegetation distribution given a prescribed climate and did not include any carbon cycle components. A next logical step was to make them dynamic, i.e. describe process-based the growth and decay of the vegetation including disturbances such as fire. This lead to the advent of "Dynamical Global Vegetation Models" (DGVM) which necessarily needed to include also dynamic carbon pools. DGVMs, however, matured only in the late 1990s and early 2000s (Brovkin et al. 1997; Sitch et al. 2003; Bonan et al. 2003; Krinner et al. 2005).

A comprehensive description of land processes including interactive carbon cycle components as a surface module within an ESM hence required to merge the three domains outlined in Fig. 3.17. Such a scheme in combination with a coupled ocean general circulation—ocean carbon cycle module constituted the ultimate goal for achieving a process-based, dynamic, fully coupled carbon cycle—climate model (CCCM). Such a model allows to conduct emission-driven simulations with internally calculated climate feedbacks on carbon uptake and redistribution within

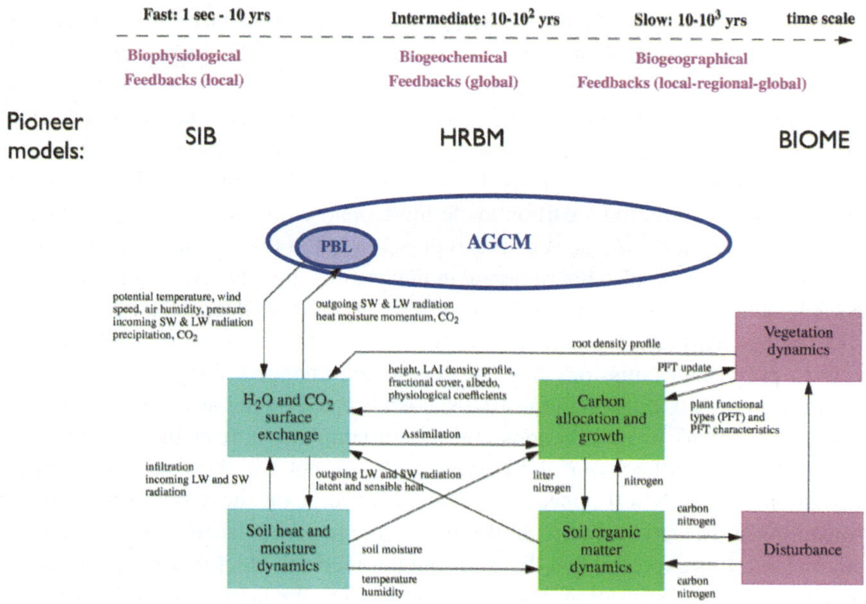


**Fig. 3.17** Essential processes in land surface modules of state-of-the-art comprehensive ESMs (Adapted from Heimann 2001)

the fast global carbon cycle (atmosphere, land, ocean). The very first group to achieve this ambitious goal was the UK Hadley center group (Cox et al. 2000).

In Hamburg, on the land side the merging of the three domains took somewhat longer. The initiative to build such a model came from the newly established Max Planck Institute for Biogeochemistry in Jena. In a joint collaboration using building blocks from the Biosphere Energy-Transfer Hydrology (BETHY) scheme (Knorr and Heimann 2001a,b) and from the Lund-Potsdam-Jena (LPJ) DGVM (Sitch et al. 2003) the "Jena Scheme for Biosphere-Atmosphere Coupling in Hamburg" (JSBACH) was created. Together with the updated ocean carbon cycle model HAMOCC5 these modules provided the essential carbon cycle components within the Hamburg ESM at the time (ECHAM5 + MPI-OM, Raddatz et al. 2007). Later, the biogeochemical cycles were synchronously coupled with terrestrial biogeophysical processes, vegetation dynamics and land use (Brovkin et al. 2009; Jungclaus et al. 2010).

## 3.10 Conclusions

Carbon cycle science and modelling has come of age. Most current state-of-the-art ESMs include now a fully coupled carbon cycle. Climate model intercomparison projects include specific work packages (e.g. the Coupled Climate-Carbon Cycle

Model Intercomparison Project, $C^4$MIP, Jones et al. 2016) targeted at evaluating the carbon cycle components, either offline of as part of the fully coupled ESM. Emission driven scenario calculations have become standard for assessing climate change projections during this century. In this long-term effort, contributions from Hamburg over the last 5 decades have been substantial, as outlined in the previous sections focusing on the time period before 2000. Hamburg pioneered the very first coupled ocean circulation—carbon cycle model and with Osnabrück contributed a first spatially explicit, climate sensitive representation of the terrestrial carbon cycle. Hamburg efforts were also instrumental in integrating atmospheric observations into ESMs leading to comprehensive Carbon Cycle Data Assimilation Systems.

Overall, at least on a centennial scale, the first order response of the global carbon cycle to prescribed emissions has not very much changed since the assessments with the simple box-models in the 1970s: corroborated by the near constancy of the "airborne fraction" in observations and model simulations over the last 60 years. So far, climate feedbacks on the carbon cycle, except for short term interannual fluctuations (e.g. ENSO) appear to be small, at least on the global scale. Model projections into a warmer future, however, diverge in particular regarding the carbon sink on land. It is expected that this sink can not increase indefinitely but the size of the land excess carbon storage capacity and whether the land sink might even turn into a source later this century is still debated (Arora et al. 2020).

Understanding, quantifying and predicting the global carbon cycle—climate system is still far from perfect. Key uncertainties include in particular still the $CO_2$ fertilisation effect of terrestrial vegetation (Walker et al. 2021) which clearly exists, but its magnitude is subject to a plethora of controlling factors: besides climate and water prominently also by available nutrients (Zaehle 2013). Also several aspects of the land carbon system are still very poorly described in current CCCMs. E.g. permafrost in high latitudes contains large carbon storage, which might become vulnerable in a warming climate (Schuur et al. 2022). And land ecosystems are described with only a few plant functional types, and might thus ignore an inherent biodiversity-related resistance to disturbances. Disturbances also impact plant mortality rates and thus the residence time of carbon in the ecosystem. Important uncertainties also exist regarding how changes in ocean circulation and stratification will impact ocean carbon uptake and modify the biological ocean carbon pumps. Most importantly, land use and land management by humans increasingly dominates terrestrial ecosystems and their carbon storage capacity. This factor crucially determines the global terrestrial land sink in the twenty-first century. Addressing this factor in global ESMs is still in its infancy and should obtain a high priority (Pongratz et al. 2009).

# References


Arora VK, Katavouta A, Williams RG, Jones CD, Brovkin V, Friedlingstein P, Schwinger J, Bopp L, Boucher O, Cadule P, Chamberlain MA, Christian JR, Delire C, Fisher RA, Hajima T, Ilyina T, Joetzjer E, Kawamiya M, Koven CD, Krasting JP, Law RM, Lawrence DM, Lenton A, Lindsay K, Pongratz J, Raddatz T, Séférian R, Tachiiri K, Tjiputra JF, Wiltshire A, Wu T, Ziehn T (2020) Carbon–concentration and carbon–climate feedbacks in CMIP6 models and their comparison to CMIP5 models. Biogeosciences 17(16):4173–4222

Arrhenius SA (1896) On the influence of carbonic acid in the air upon the pemperature of the ground. Philos Mag J Sci 41(251):237–276

Bacastow R, Maier-Reimer E (1990) Ocean-circulation model of the carbon cycle. Clim Dyn 4(2):95–125

Bacastow R, Maier-Reimer E (1991) Dissolved organic carbon in modeling oceanic new production. Glob Biogeochem Cycles 5(1):71–85

Barnola JM, Raynaud D, Korotkevich YS, Lorius C (1987) Vostok ice core provides 160,000-year record of atmospheric $CO_2$. Nature 329(6138):408–414

Bennett BF, Salawitch RJ, McBride LA, Hope AP, Tribett WR (2024) Quantification of the airborne fraction of atmospheric $CO_2$ reveals stability in global carbon sinks over the past six decades. J Geophys Res Biogeosci 129(3), e2023JG007760

Berner W, Oeschger H, Stauffer B (1980) Information on the $CO_2$ cycle from ice core studies. Radiocarbon 22(2):227–235

Bolin B (1979) On the role of the atmosphere in biogeochemical cycles. Q J R Meteor Soc 105:25–42

Bolin B (ed) (1981) Carbon cycle modelling (SCOPE 16). Wiley. https://scope.dge.carnegiescience.edu/SCOPE_16/SCOPE_16.html. ISBN: 0 471 10051 X

Bolin B, Keeling CD (1963) Large-scale atmospheric mixing as deduced from the seasonal and meridional variation of atmospheric carbon dioxide. J Geophys Res Atmos 68(13):3899–3920

Bolin B, Degens ET, Kempe S, Ketner P (eds) (1979) SCOPE 13 - The global carbon cycle. Wiley. ISBN: 9780471997108. https://web.archive.org/web/20140518205209/http://www.scopenvironment.org/downloadpubs/scope13/contents.html

Bonan GB, Levis S, Sitch S, Vertenstein M, Oleson KW (2003) A dynamic global vegetation model for use with climate models: concepts and description of simulated vegetation dynamics. Glob Change Biol 9(11):1543–1566

Bovensmann H, Burrows JP, Buchwitz M, Frerick J, Noël S, Rozanov VV, Chance KV, Goede APH (1999) SCIAMACHY: Mission objectives and measurement modes. J Atmos Sci 56(2):127–150

Brook EJ (2008) Palaeoclimate windows on the greenhouse. Nature 453(7193):291–292

Brovkin V, Ganopolski A, Svirezhev Y (1997) A continuous climate-vegetation classification for use in climate-biosphere studies. Ecol Model 101(2–3):251–261

Brovkin V, Raddatz T, Reick CH, Claussen M, Gayler V (2009) Global biogeophysical interactions between forest and climate. Geophys Res Lett 36:L07405. https://doi.org/10.1029/2009GL037543

Callendar G (1938) The artificial production of carbon dioxide and its influence on temperature. Q J R Meteor Soc 64:223–240

Ciais P, Sabine C, Bala G, Bopp L, Brovkin V, Canadell J, Chhabra A, DeFries R, Galloway J, Heimann M, Jones C, Le Quèrè C, Myneni RB, Piao S, Thornton P (2014) Carbon and other biogeochemical cycles. In: Stocker T, Qin D, Plattner GK, Tingor M, Allen S, Boshung J, Nauels A, Xia Y, Bex V, Midgley P (eds) Climate change 2013 – the physical science basis: working group I Contribution to the fifth assessment report of the intergovernmental panel on climate change. Cambridge University Press, pp 465–570. https://doi.org/10.1017/CBO9781107415324.015

Cox P, Betts R, Jones C, Spall S, Totterdell I (2000) Acceleration of global warming due to carbon-cycle feedbacks in a coupled climate model. Nature 408(6809):184–187

Crisp D, Atlas RM, Breon F, Brown L, Burrows J, Ciais P, Connor B, Doney S, Fung IY, Jacob DJ, Miller C, O'Brien D, Pawson S, Randerson J, Rayner PJ, Salawitch R, Sander S, Sen B, Stephens G, Tans P, Toon G, Wennberg P, Wofsy S, Yung Y, Kuang Z, Chudasama B, Sprague G, Weiss B, Pollock R, Kenyon D, Schroll S (2004) The orbiting carbon observatory (OCO) mission. Adv Space Res-Series 34(4):700–709

Delmas R, Ascencio J, Legrand M (1980) Polar ice evidence that atmospheric $CO_2$ 20,000yr BP was 50% of present. Nature 284(5752):155–157

Denning AS, Holzer M, Gurney K, Heimann M, Law R, Rayner PJ, Fung IY, Fan S, Taguchi S, Friedlingstein P, Balkanski Y, Taylor J, Maiss M, Levin I (1999) Three-dimensional transport and concentration of $SF_6$ - A model intercomparison study (TransCom 2). Tellus B 51(2):266–297

Enting I, Wigley T, Heimann M (1994) Future emissions and concentrations of carbon dioxide: Key ocean/atmosphere/land analyses, vol 31. CSIRO Division of Atmospheric Research, Melbourne

Esser G (1984) The significance of biospheric carbon pools and fluxes for the atmospheric $CO_2$: A proposed model structure. Prog Biometeorol 3:253–294

Esser G (1987) Sensitivity of global carbon pools and fluxes to human and potential climatic impacts. Tellus B 39B(3):245–260

Esser G (1991) Osnabrück Biosphere Model: structure, construction, results. In: Esser G, Overdieck D (eds) Modern ecology: basic and applied aspects. Elsevier, Amsterdam, pp 679–709

Esser G (1995) Contribution of monsoon Asia to the carbon budget of the biosphere, past and future. Vegetatio 121:175–188

Esser G, Lautenschlager M (1994) Estimating the change of carbon in the terrestrial biosphere from 18'000 BP to present using a carbon cycle model. Environ Pollut 83(1–2):45–53

Fan S, Gloor M, Mahlman J, Pacala S, Sarmiento J, Takahashi T, Tans P (1998) A large terrestrial carbon sink in North America implied by atmospheric and oceanic carbon dioxide data and models. Science 282(5388):442–446

Farquhar G, von Caemmerer S, Berry J (1980) A biochemical model of photosynthetic $CO_2$ assimilation in leaves of C3 species. Planta 149(1):78–90

Foote E (1856) Circumstances affecting the heat for the sun's rays. Am J Sci Arts 22:382–383

Fourier JBJ (1837) General remarks on the temperature of the earth and outer space. Am J Sci 32:1–20

Friedlingstein P, Dufresne J, Cox P, Rayner PJ (2003) How positive is the feedback between climate change and the carbon cycle. Tellus 55(2):692–700

Friedlingstein P, Cox P, Betts RA, Bopp L, von W Bloh, Brovkin V, Cadule P, Doney S, Eby M, Fung I (2006) Climate–carbon cycle feedback analysis: Results from the C4MIP model intercomparison. J Clim 19(14):3337–3353

Fung I, Prentice K, Matthews E, Lerner J, Russell G (1983) 3-dimensional tracer model study of atmospheric $CO_2$ - response to seasonal exchanges with the terrestrial biosphere. J Geophys Res 88(NC2):1281–1294

GARP Joint Organizing Committee (1975) The physical basis of climate and climate modelling. GARP Publication Series 16

Garratt JR, Pearman GI (1973) Large-scale $CO_2$ fluxes in the southern hemisphere troposphere. Nat Phys Sci 242(117):54–56

Giering R, Kaminski T (1998) Recipes for adjoint code construction. ACM Trans Math Softw 24(4):437–474

Hasselmann K (1979a) On the problem of multiple time scales in climate modeling. In: Developments in atmospheric science: man's impact on climate. Elsevier, pp 43–55

Heimann M (2001) Modeling and evaluating terrestrial biospheric exchanges of water, carbon dioxide, and oxygen in the global climate system. In: Bengtsson LO, Hammer CU (eds) Geosphere-biosphere interactions and climate. Cambridge University Press, Cambridge, pp 52–65

Heimann M, Keeling CD (1989) A three-dimensional model of atmospheric $CO_2$ transport based on observed winds: 2. Model description and simulated tracer experiments. AGU Geophys Monogr 55:237–275

Heimann M, Keeling CD, Fung IY (1986) Simulating the atmospheric carbon dioxide distribution with a three-dimensional tracer model. In: Trabalka JR, Reichle DE (eds) The changing carbon cycle: a global analysis. Springer, New York, pp 16–49

Heimann M, Esser G, Haxeltine A, Kaduk J, Kicklighter DW, Knorr W, Kohlmaier GH, McGuire AD, Melillo J, Moore B III, Otto RD, Prentice IC, Sauf W, Schloss A, Sitch S, Wittenberg U, Würth G (1998) Evaluation of terrestrial carbon cycle models through simulations of the seasonal cycle of atmospheric $CO_2$: First results of a model intercomparison study. Glob Biogeochem Cycles 12(1):1–24

Heinze C, Maier-Reimer E, Winn K (1991) Glacial $pCO_2$ reduction by the world ocean: Experiments with the Hamburg carbon cycle model. Paleoceanography 6(4):395–430

Hooss G, Voss R, Hasselmann K, Maier-Reimer E, Joos F (2001) A nonlinear impulse response model of the coupled carbon cycle-climate system (NICCS). Clim Dyn 18(3–4):189–202

Jones CD, Arora V, Friedlingstein P, Bopp L, Brovkin V, Dunne J, Graven H, Hoffman F, Ilyina T, John JG, Jung M, Kawamiya M, Koven C, Pongratz J, Raddatz T, Randerson JT, Zaehle S (2016) C4MIP – The coupled climate–carbon cycle model intercomparison project: experimental protocol for CMIP6. Geosci Model Dev 9(8):2853–2880

Joos F, Bruno M, Fink R, Siegenthaler U, Stocker T, LeQuéré C (1996) An efficient and accurate representation of complex oceanic and biospheric models of anthropogenic carbon uptake. Tellus B 48(3):397–417

Joos F, Roth R, Fuglestvedt JS, Peters GP, Enting IG, von W Bloh, Brovkin V, Burke EJ, Eby M, Edwards NR, Friedrich T, Frölicher TL, Halloran PR, Holden PB, Jones C, Kleinen T, Mackenzie FT, Matsumoto K, Meinshausen M, Plattner GK, Reisinger A, Segschneider J, Shaffer G, Steinacher M, Strassmann K, Tanaka K, Timmermann A, Weaver AJ (2013) Carbon dioxide and climate impulse response functions for the computation of greenhouse gas metrics: a multi-model analysis. Atmos Chem Phys 13(5):2793–2825

Jungclaus JH, Lorenz SJ, Timmreck C, Reick CH, Brovkin V, Six K, Segschneider J, Giorgetta MA, Crowley TJ, Pongratz J, Krivova NA, Vieira LE, Solanki SK, Klocke D, Botzet M, Esch M, Gayler V, Haak H, Raddatz TJ, Roeckner E, Schnur R, Widmann H, Claussen M, Stevens B, Marotzke J (2010) Climate and carbon-cycle variability over the last millennium. Clim Past 6(5):723–737

Junge CE, Czeplak G (1968) Some aspects of the seasonal variation of carbon dioxide and ozone. Tellus A Dyn Meteorol Oceanogr 20(3):422

Kaminski T, Heimann M (2001) Inverse modeling of atmospheric carbon dioxide fluxes. Science 294(5541):259–259

Kaminski T, Heimann M, Giering R (1999a) A coarse grid three-dimensional global inverse model of the atmospheric transport. I- adjoint model and jacobian matrix. J Geophys Res 104(D15):18555

Kaminski T, Heimann M, Giering R (1999b) A coarse grid threedimensional global inverse model of the atmospheric transport, 2, inversion of the transport of co in the 1980s. J Geophys Res 104(D15):18555–18581

Kaminski T, Rayner PJ, Heimann M, Enting I (2001) On aggregation errors in atmospheric transport inversions. J Geophys Res-Atmos 106(D5):4703–4715

Keeling CD (1960) The concentration and isotopic abundances of carbon dioxide in the atmosphere. Tellus 12(2):200–203

Keeling CD (1973) The carbon dioxide cycle: Reservoir models to depict the exchange of atmospheric carbon dioxide with the oceans and land plants. In: Rasool SI (ed) Chemistry of the lower atmosphere. Plenum Publishing Company, New York, pp 251–329

Keeling R (2022) Scripps $CO_2$ program. https://scrippsco2.ucsd.edu. Accessed: Nov 2022

Keeling CD, Piper S, Heimann M (1989) A three-dimensional model of atmospheric $CO_2$ transport based on observed winds: 4. mean annual gradients and interannual variations. AGU Geophys Monogr 55:305–363

Knorr W, Heimann M (2001a) Uncertainties in global terrestrial biosphere modeling: 1. A comprehensive sensitivity analysis with a new photosynthesis and energy balance scheme. Global Biogeochem Cycles 15(1):207–225

Knorr W, Heimann M (2001b) Uncertainties in global terrestrial biosphere modeling, Part II: Global constraints for a process-based vegetation model. Global Biogeochem Cycles 15(1):227–246

Komhyr WD, Gammon RH, Harris TB, Waterman LS, Conway TJ, Taylor WR, Thoning KW (1985) Global atmospheric $CO_2$ distribution and variations from 1968–1982 NOAA/GMCC $CO_2$ flask sample data. J Geophys Res Atmos 90(D3):5567–5596

Krinner G, Viovy N, de N Noblet-Ducoudré, Ogee J, Polcher J, Friedlingstein P, Ciais P, Sitch S, Prentice IC (2005) A dynamic global vegetation model for studies of the coupled atmosphere-biosphere system. Global Biogeochem Cycles 19(1):GB1015. https://doi.org/10.1029/2003GB002199

Law RM, Rayner PJ, Denning AS, Erickson D, Fung IY, Heimann M, Piper SC, Ramonet M, Taguchi S, Taylor JA, Trudinger CM, Watterson IG (1996) Variations in modeled atmospheric transport of carbon dioxide and the consequences for $CO_2$ inversions. Global Biogeochemical Cycles 10(4):783–796

Lieth H (1975) Modeling the primary productivity of the world. In: Lieth, Whittaker (eds) Primary productivity of the biosphere, Ecological studies 14. Springer, New York, pp 237–283

Lieth H, Esser G (1985) The attempt to simulate the global carbon flux from 1860 to 1981 using the Osnabrück Biosphere Model. In: Mitt. Geol.-Paläontolog. Inst. Univ. Hamburg, SCOPE/UNEP Sonderband, vol 58. University of Hamburg, Hamburg, pp 137–144

Lüthi D, Le M Floch, Bereiter B, Blunier T, Barnola JM, Siegenthaler U, Raynaud D, Jouzel J, Fischer H, Kawamura K, Stocker TF (2008) High-resolution carbon dioxide concentration record 650,000–800,000 years before present. Nature 453(7193):379–382

Maier-Reimer E, Bacastow R (1990) Modelling of geochemical tracers in the ocean. In: Schlesinger M (ed) Climate-Ocean Interaction. Kluwer Academic, Boston, pp 233–267

Maier-Reimer E, Hasselmann K (1987) Transport and storage of $CO_2$ in the ocean - an inorganic ocean-circulation carbon cycle model. Clim Dyn 2(2):63–90

Maier-Reimer E, Mikolajewicz U, Winguth A (1996) Future ocean uptake of $CO_2$ interaction between ocean circulation and biology. Clim Dyn 12(10):711–722

Müller J, Joos F (2020) Global peatland area and carbon dynamics from the last glacial maximum to the present – a process-based model investigation. Biogeosciences 17(21):5285–5308

Oeschger H, Heimann M (1983) Uncertainties of predictions of future atmospheric $CO_2$ concentrations. J Geophys Res Oceans 88(NC2):1258–1262

Oeschger H, Siegenthaler U, Schotterer U, Gugelmann A (1975) A box diffusion model to study the carbondioxide exchange in nature. Tellus B 17:168–192

Plass GN (1956) The carbon dioxide theory of climatic change. Tellus 8(2):140–154

Pongratz J, Reick CH, Raddatz T, Claussen M (2009) Effects of anthropogenic land cover change on the carbon cycle of the last millennium. Global Biogeochem Cycles 23(4):189

Prentice IC, Cramer W, Harrison SP, Leemans R, Monserud RA, Solomon AM (1992) A global biome model based on plant physiology and dominance, soil properties and climate. J Biogeogr 19, 117–134

Raddatz TJ, Reick CH, Knorr W, Kattge J, Roeckner E, Schnur R, Schnitzler KG, Wetzel P, Jungclaus J (2007) Will the tropical land biosphere dominate the climate–carbon cycle feedback during the twenty-first century. Clim Dyn 29(6):565–574

Rayner PJ (2001) Atmospheric perspectives on the ocean carbon cycle. In: Schulze ED, Heimann M, Harrison S, Holland E, Lloyd J, Prentice IC, Schimel D (eds) Global biogeochemical cycles in the climate system. Academic Press, San Diego, pp 285–294

Rayner PJ, O'Brien DM (2001) The utility of remotely sensed $CO_2$ concentration data in surface source inversions. Geophys Res Lett 28(1):175–178

Rayner PJ, Scholze M, Knorr W, Kaminski T, Giering R, Widmann H (2005) Two decades of terrestrial carbon fluxes from a carbon cycle data assimilation system (CCDAS). Global Biogeochem Cycles 19(2):n/a

Revelle R, Suess HE (1957) Carbon dioxide exchange between atmosphere and ocean and the question of an increase of atmospheric $CO_2$ during the past decades. Tellus 9(1):18–27

Roedenbeck C, Zaehle S, Keeling R, Heimann M (2018a) History of El Nino impacts on the global carbon cycle 1957-2017: a quantification from atmospheric $CO_2$ data. Philos Trans R Soc B Biol Sci 373(1760), 20170303

Roedenbeck C, Zaehle S, Keeling R, Heimann M (2018b) How does the terrestrial carbon exchange respond to inter-annual climatic variations? A quantification based on atmospheric $CO_2$ data. Biogeosciences 15(8):2481–2498

Roedenbeck C, Zaehle S, Keeling R, Heimann M (2020) The european carbon cycle response to heat and drought as seen from atmospheric $co_2$ data for 1999-2018. Philos Trans R Soc Lond B Biol Sci 375(1810):20190506

Schimel D, Alves AC De Oliveira, Enting IG, Heimann M, Joos F, Raynaud D, Wigley T, Prather M, Derwent RG, Ehhalt DH, Fraser P, Sanhueza E, Zhou XH, Jonas P, Charlson R, Rohde H, Sadasivan S, Shine KP, Fouquart Y, Ramaswamy V, Solomon S, Srinivasan J, Albritton D, Isaksen ISA, Lal M, Wuebbles D (1996) Radiative forcing of climate change. In: Hougthon JT, Meira LG Filho, Callander BA, Kattenberg A, Maskell K (eds) Climate change 1995: the science of climate change. Cambridge University Press, Cambridge, pp 65–131

Schuermann GJ, Kaminski T, Koestler C, Carvalhais N, Vossbeck M, Kattge J, Giering R, Roedenbeck C, Heimann M, Zaehle S (2016) Constraining a land-surface model with multiple observations by application of the MPI-Carbon Cycle Data Assimilation System v1.0. Geosci Model Dev 9:2999–3026

Schuur EA, Abbott BW, Commane R, Ernakovich J, Euskirchen E, Hugelius G, Grosse G, Jones M, Koven C, Leshyk V, Lawrence D, Loranty MM, Mauritz M, Olefeldt D, Natali S, Rodenhizer H, Salmon V, Schädel C, Strauss J, Treat C, Turetsky M (2022) Permafrost and climate change: Carbon cycle feedbacks from the warming Arctic. Annu Rev Environ Resour 47(1):343–371

Sellers P, Randall D, Collatz G, Berry J, Field C, Dazlich D, Zhang C, Collelo G, Bounoua L (1996a) A revised land surface parameterization (SiB2) for atmospheric GCMs. Part I: Model formulation. J Clim 9(4):676–705

Sellers PJ, Tucker CJ, Collatz GJ, Los SO, Justice CO, Dazlich DA, Randall DA (1996b) A revised land surface parameterization (SiB2) for atmospheric GCMs. Part II: The generation of global fields of terrestrial biophysical parameters from satellite data. J Clim 9(4):706–737

Sellers PJ, Dickinson RE, Randall DA, Betts AK, Hall FG, Berry JA, Collatz GJ, Denning AS, Mooney HA, Nobre CA, Sato N, Field CB, Henderson-Sellers A (1997) Modeling the exchanges of energy, water, and carbon between continents and the atmosphere. Science 275(5299):502–509

Shine KP, Derwent RG, Wuebbles D, Morcrette JJ (1990) Radiative forcing of climate. In: Hougthon JT, Jenkins GJ, Ephraums JJ (eds) Climate change: The IPCC scientific assessment. Cambridge University Press, Cambridge, pp 41–68

Sitch S, Smith B, Prentice I, Arneth A, Bondeau A, Cramer W, Kaplan JO, Levis S, Lucht W, Sykes M, Thonicke K, Venevsky S (2003) Evaluation of ecosystem dynamics, plant geography and terrestrial carbon cycling in the LPJ dynamic global vegetation model. Global Change Biol 9(2):161–185

Stephens BB, Keeling RF, Heimann M, Six KD, Murnane R, Caldeira K (1998) Testing global ocean carbon cycle models using measurements of atmospheric $O_2$ and $CO_2$ concentration. Global Biogeochem Cycles 12(2):213–230

Tans P, Fung IY, Takahashi T (1990) Observational contrains on the global atmospheric $CO_2$ budget. Science 247(4949):1431–1438

Tyndall J (1861) The Bakerian Lecture: On the absorption and radiation of heat by gases and vapours, and on the physical connexion of radiation, absorption, and conduction. Philos Trans R Soc Lond 151:1–36

Walker AP, De Kauwe MG, Bastos A, Belmecheri S, Georgiou K, Keeling RF, McMahon SM, Medlyn BE, Moore DJP, Norby RJ, Zaehle S, Anderson-Teixeira KJ, Battipaglia G, Brienen RJW, Cabugao KG, Cailleret M, Campbell E, Canadell JG, Ciais P, Craig ME, Ellsworth DS, Farquhar GD, Fatichi S, Fisher JB, Frank DC, Graven H, Gu L, Haverd V, Heilman K, Heimann M, Hungate BA, Iversen CM, Joos F, Jiang M, Keenan TF, Knauer J, Körner C, Leshyk VO, Leuzinger S, Liu Y, MacBean N, Malhi Y, McVicar TR, Penuelas J, Pongratz J, Powell AS, Riutta T, Sabot MEB, Schleucher J, Sitch S, Smith WK, Sulman B, Taylor B, Terrer C, Torn MS, Treseder KK, Trugman AT, Trumbore SE, Mantgem PJ, Voelker SL, Whelan ME, Zuidema PA (2021) Integrating the evidence for a terrestrial carbon sink caused by increasing atmospheric $CO_2$. New Phytologist 229:2413

Woodwell G, Whittaker R, Reiners W, Likens G, Delwiche C, Botkin D (1978) The biota and the world carbon budget. Science 199(4325):141–146

Yokota T, Oguma H, Morino I, Higurashi A, Aoki T, Inoue G (2004) Test measurements by a BBM of the nadir-looking SWIR FTS aboard GOSAT to monitor $CO_2$ column density from space. In: Tsay SC, Yokota T, Ahn MH (eds) Passive optical remote sensing of the atmosphere and clouds IV, SPIE, p 19

Zaehle S (2013) Terrestrial nitrogen - carbon cycle interactions at the global scale. Philos Trans R Soc B-Biol Sci 368:20130125. https://doi.org/10.1098/rstb.2013.0125

# Chapter 4
# Multiple Equilibria in Climate Models

**Martin Claussen, Hans von Storch, Martin Heimann,
Robert Sausen, and Eduardo Zorita**

**Abstract** In the 1960s and 1970s, the dynamics of the global climate system were investigated using models that exhibited a high degree of idealization. In his groundbreaking article in Tellus in 1976, Klaus Hasselmann raised the question of whether these models only showed abrupt transitions because they oversimplified the dynamics of the climate system and did not reflect natural climate variability. In this chapter, Hasselmann's assumption is therefore re-examined. To this end, three examples of multiple equilibria and rapid transitions in the climate system are discussed: Snowball Earth, the Atlantic meridional overturning circulation and the interaction between climate and vegetation. The discussion of these examples begins with predictions from idealized models, followed by evidence of abrupt transitions found in palaeoclimatic proxies that stimulated the search for unstable feedbacks using complex models. Discussion of these examples start with predictions from idealized models, followed by evidence of abrupt transitions found in paleo climate proxies which fostered simulation of unstable feedbacks using complex models. Climate scientists who have worked in Hamburg have made significant contributions to all of these examples.

## 4.1 Unstable Feedback Models - Just an Artefact?

In his pioneering paper on stochastic climate models, Klaus Hasselmann (1976) wrote: *A basic difficulty of unstable feedback models (apart from - or possibly because of – their high degree of idealisation) is that they tend to predict climatic variations as flip-flop transitions and therefore fail to reproduce the observed continuous spectrum of climatic variability.* At the time when Hasselmann wrote this statement, most models used to study the dynamics of the global climate system were highly simplified, so-called energy balance models (see below). These models did not exhibit any free climate variability unless some stochastic noise was prescribed as a substitute of weather variability. Would then, in turn, complex models of a low degree idealisation, also called *quasi-realistic models*, which predict a more or less realistic weather and climate variability, not show any flip-flop transitions or multiple equilibrium solutions for the same climate state? This

 
H. von Storch et al., *Climate Science Concepts Born in Hamburg*,
https://doi.org/10.1007/978-3-031-81108-1_4

question, which has kept climate modellers busy ever since (Valdes 2011), will be explored in this chapter.

At the end of the 1970s and early 1980s, Hasselmann pursued the development of complex, quasi-realistic models in his institute, jointly with Günter Fischer, professor of theoretical meteorology and Jürgen Sündermann, professor of oceanography at the Universität Hamburg (see Chap. 1).[1] Some contributions to the topic of multiple equilibria and flip-flop transitions in complex models, which had some impact in the international science community, came from scientists working in Hamburg, and these will be highlighted.

The chapter focuses on three cases. First, multiple equilibria in relation to extreme global temperature transitions that lead to a flip-flop between a fully glaciated Earth and an Earth without any ice sheets are discussed. The second example highlights multiple equilibria and rapid transitions in oceanic circulation, namely the Atlantic Overturning Circulation (AMOC). And the third example focuses on multiple equilibria in relation to climate-vegetation interaction. All examples follow the same structure; they start with predictions by using models of low complexity, briefly discuss palaeo-climatic evidence for flip-flop transitions in nature and then focus on first attempts of using quasi-realistic models. For each example, a brief outline of how the story continues is given at the end.

## 4.2 Snowball Earths

Likely independent of each other, Michail Budyko (1969) and William Sellers (1969) set up highly idealized models of the energy balance of the Earth's climate to understand the origin of the Quaternary glacial cycles. In their models, the temperature $T$ averaged over a zonal belt between two latitudes was computed based on the conservation of energy by:

$$c\frac{\partial T}{\partial t} = Q(1 - \alpha(T)) - I(T) + A(T - T_p) \tag{4.1}$$

where $c$ is the specific heat capacity of the relevant climate components, $Q$ is the net, annual mean insolation at a given latitude. $\alpha$ is the albedo, i.e. the ratio of incoming to outgoing solar radiation at the top of the atmosphere. $I$ is the outgoing heat radiation of the surface and the atmosphere, and $A$ refers to the convergence of the meridional heat flux between two latitudinal belts. The albedo $\alpha$ is assumed to depend on the temperature $T$; at a high temperature, i.e. in a warm latitudinal

[1] On a personal note, when I left my postdoc position in Klaus Hasselmann's group in 1985, Klaus Hasselmann asked me about my research plans. I answered that I would try to apply this idea of stochastic climate models to atmospheric boundary flow over randomly varying heterogeneous surfaces. Klaus Hasselmann suggested that I should rather go for complex modelling to more realistically tackle the problem - a piece of advice which I followed.

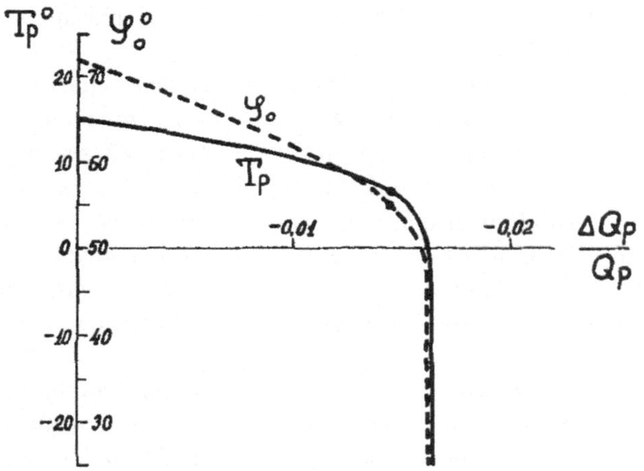


**Fig. 4.1** Dependence of the Earth's temperature $T_p$ and ice-cover boundary $\phi_o$ on small, negative variations $\Delta Q_p$ of planetary insolation $Q_p$. With decreasing insolation, the planetary temperature $T_p$ (vertical axis, left scale in degree Celsius) decreases and the ice-cover boundary $\phi_o$ (vertical axis, right scale in degrees northern latitude) shifts south until a threshold in $\Delta Q_p/Q_p$ is crossed and the temperature rapidly drops and the ice-cover boundary rushes toward the equator. The figure is reprinted from Budyko (1969) with the permission of *Tellus A*

belt, the Earth's surface is assumed to be ice-free. Hence the albedo is low. In turn, at a low temperature, the albedo is high, and the incoming solar radiation is more strongly reflected than in a latitudinal belt of low albedo. This dependence $\alpha(T)$ is the crucial ice-albedo effect which triggers a highly non-linear behaviour of the model system. For simplicity, the advection term $A$ of heat is assumed to be a function of local temperature $T$ and planetary, or global mean temperature $T_p$: $A \sim (T - T_p)$; if a latitudinal belt has a temperature warmer than the global average, then heat is transported to other latitudinal belts, and *vice versa*. On a global average, the advection of heat between latitudinal belts vanishes. The stationary planetary temperature $T_p$ is then a function of globally averaged insolation, albedo and heat radiation.

$$0 = Q_p(1 - \alpha(T_p)) - I(T_p) \quad (4.2)$$

Budyko (1969) and Sellers (1969) found a critical threshold - now more catchy called a *tipping point*[2] - in insolation around $\Delta Q_p/Q_p \approx -0.015$ beyond which the Earth would completely freeze (see Fig. 4.1). Further analysis reveals that because of the non-linear ice-albedo feedback, multiple equilibrium states of the planetary

[2] The term tipping point was coined in the late 1950s in sociological studies. The tipping point in using the term was the publication of Malcolm Gladwell's book "The tipping point - when little things make a big difference". The term entered climate science in the early 2000s (Van Nes et al. 2016)

temperature $T_p$ occur between which $T_p$ switches when increasing or decreasing $Q$ crosses certain thresholds.

Back then, Budyko (1969) called this behaviour a *climate catastrophe*. He wrote: *From such a viewpoint the Quaternary History of the Earth seems to be the period of coming climatic catastrophe due to which the existence of higher forms of organic life on our planet may be exterminated.* On the other hand, Budyko and Sellers were well aware of a possible interference of humans with the climate system. Sellers (1969) noted that *Thus, man's activity, if it continues unabated, should eventually lead to the elimination of the ice caps ... it is impossible to state how long it will take for this warming to occur – possibly as little as 100 years or as long as 1000 years.*

The term *climate catastrophe* is generally, and certainly in the media, meant a catastrophe for humankind. In science, the term refers to the mathematical theory of catastrophes being developed in the 1960s by René Thom, Vladimir Arnold and others. Klaus Fraedrich, professor of theoretical meteorology in Berlin and, later, in Hamburg further explored this type of rapid changes in the Earth's energy balance by allowing the heat radiation $I(T, CO_2)$ to depend not only on temperature $T$ and, hence, a varying greenhouse effect due to water - vapour which implicitly depend on the temperature -, but also on the $CO_2$ concentration in the atmosphere, the *dry greenhouse effect* (Fraedrich 1979)

Peter Lemke, a doctoral student in Hasselmann's department, used the Budyko-Sellers type of energy balance models to investigate the effect of weather noise on the spectrum of decadal to multi-millennial temperature variability (Lemke 1977) (see also Sect. 2.2.2). He added a term with pre-scribed stochastic noise $S'$ to the energy balance:

$$c\frac{\partial T}{\partial t} = \langle Q(1-\alpha(T))\rangle - \langle I(T)\rangle + \langle A(T, T_p)\rangle + S' \quad (4.3)$$

Thereby, the temperature $T$ is not a deterministic variable as in the models of Budyko and Sellers, but a randomly varying variable. Lemke (1977) noted that his stochastic energy balance model could explain the reconstructed spectra of $T$. Moreover, he found that his model reveals a similar sensitivity of $T_p$ to small perturbations $\Delta Q_p$ in insolation, as seen in the models of Budyko and Sellers. In particular, there was a critical value of the ice-albedo feedback beyond which the spectral response becomes infinite. This was the first hint on flip-flop transitions not just being an artefact of the deterministic energy balance models. However, as the stochastic weather noise had to be prescribed, the jury was still out on whether Lemke's model was still too simple to exclude the flip-flop transition as a model artefact.

Based on an idea of Grath Paltridge (1975), Hartmut Grassl (1981) added complexity to the energy balance models by relating the advection term $A$ in Eq. (4.1) to the entropy change $Q/T_{sun}(1-\alpha) + I/T$ in the climate system ($T_{sun}$ is the temperature of the Sun's photosphere). Grassl noted that *the resulting sensitivities partly agree and disagree with those of similar energy balance climate models. Disagreement is particularly high if an ice-albedo feedback is included. The feedback is strongly reduced because of opposing effects of cloud amount in high latitudes.* This statement would suggest that flip-flop transitions might vanish with increased complexity of models.

The first attempt to study the sensitivity of climate to insolation in a complex numerical model was undertaken by Richard Wetherald and Syukuro Manabe in 1975 (Wetherald and Manabe 1975). The numerical model covered a limited area of 60° of longitude using periodic boundary conditions and a highly idealised topography. No ocean heat flux was included, and the cloud cover was fixed. Wetherald and Manabe detected a high sensitivity of the modelled hydrological cycle to a small perturbation in insolation, but they did not note a rapid transition to a very cold Earth even when lowering the insolation by $\Delta Q_p/Q_p \approx -0.4$, a much larger perturbation than the critical threshold in insolation found by Budyko (1969). However, Wetherald and Manabe (1975) explored the dynamics of a *White Earth solution*. They started their model with a high planetary albedo of $\alpha_p = 0.7$ and an isothermal atmosphere at a temperature $T = 222$ K. They saw that their model stayed at a very low temperature and concluded that the White Earth solution could be an alternative climate state under present-day insolation.

Research shifted into a higher gear when geologists published clear evidence of a global glaciation of the Earth, the so-called *Snowball Earth* (e.g., Kirschvink 1992; Hoffman and Schrag 2002). Today, we know that several hundred million years ago, the Earth was completely, or almost completely, covered by ice and snow at least twice between 710 and 660 million years ago and between 645 and 635 million years ago. The older and longer Snowball phase is called *Sturtian* and the later, shorter Snowball phase, *Marinoan*, after the Sturt Gorge and the Marino Rocks near Adelaide, South Australia, where evidence of glaciation of tropical regions was detected first. For a recent review see Hoffman et al. (2017).

The most interesting feature in the geological archives appeared to be the sharp boundaries between an extremely cold and an extremely hot climate. Obviously, extreme and geologically abrupt flip-flop transitions in planetary temperature were not only a theoretical construct built on simple climate models, but likely happened in reality. The flip-flop transition was attributed to an interplay of weathering of $CO_2$ on silicate rocks and outgassing of $CO_2$ due to volcanic activity. It was assumed that a change in continental configuration, presumably the break-up of the early supercontinent Rodinia into smaller continents located near the equator, could have increased the weathering of $CO_2$ such that the $CO_2$ concentration would have drastically decreased. This perturbation in the Earth's climate energy budget could have initiated a rapid transition into a Snowball Earth within a few thousand years - in comparison with the time scale of roughly a hundred thousand years of $CO_2$ weathering, certainly abrupt. Once completely glaciated, the Earth's hydrological cycle should have been broken down, and any weathering of $CO_2$ should have stopped. Only the steady outgassing of $CO_2$ from the Earth's interior would have continued such that eventually, after several millions of years, the Snowball Earth was to melt within a few thousand years because of a super-greenhouse effect due to the extremely high $CO_2$ concentration in the atmosphere.

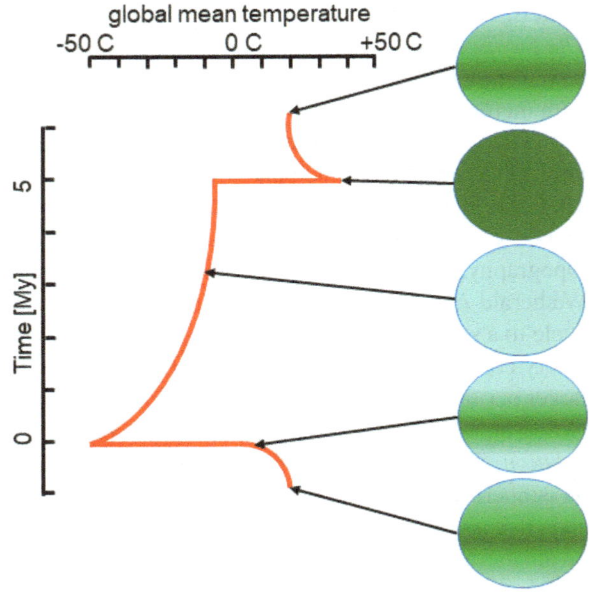


**Fig. 4.2** Sketch of the results from a deterministic, zero-dimensional energy balance model, i.e. a model without climate noise, coupled to a simple global carbon cycle model simulating the temperature of a Snowball Earth (red curve, time running upward). The ice cover is sketched on the right-hand side; a full green sphere indicates an ice-free Earth, and a light blue sphere, a fully glaciated Earth. Partial glaciation is indicated by green-blue stripes. The sketch is adapted from Hoffman and Schrag (2002) with modifications

This geological concept can easily be visualized by defining a simple dynamical system in which the change in atmospheric $CO_2$ concentration in the air depends on a (for simplicity) constant volcanic outgassing and a carbon flux due to weathering on silicate rocks, which depends on $CO_2$ and the planetary temperature. The latter, in turn, is determined by the Earth's energy balance as defined in Eq. (4.2). Again, the planetary albedo is assumed to depend on planetary temperature $\alpha_p = \alpha_p(T_P)$ and the outgoing heat radiation $I = I(T_p, CO_2)$ is a function of the temperature $T_p$ and the $CO_2$ concentration just as in Fraedrich's model (Fraedrich 1979). If properly tuned, this simple model produces Snowball Earth flip-flop transitions as shown in Fig. 4.2. But the results of the simulations, of course, depend on the numerous parameters, mainly the prescribed albedo parameterization. Therefore, this simple model is more a check of concept rather than any realistic description of the Sturtian and Marinoan Snowball Earths.

By using models of intermediate complexity, either zonally symmetric models (Stone and Yao 2004) or the 2-dimensional model CLIMBER (Donnadieu et al. 2004) or a simplified Earth system model (Lucarini et al. 2010) further attempts have been made on studying the dynamics or the equilibria of the Snowball Earth. Marotzke and Botzet (2007) were the first to test the sensitivity of a quasi-realistic model to extreme changes in the solar constant by using an early version of the MPI-ESM (Max Planck Institute - Earth System Model). Marotzke and Botzet (2007) found that when switching off the solar energy flux almost completely, the model predicts a rapid transition into a state with globally subfreezing surface temperatures and the global ocean completely covered by sea ice within some 19 years. That is, of course, not a surprise, but a proof of a well-functioning climate system model that

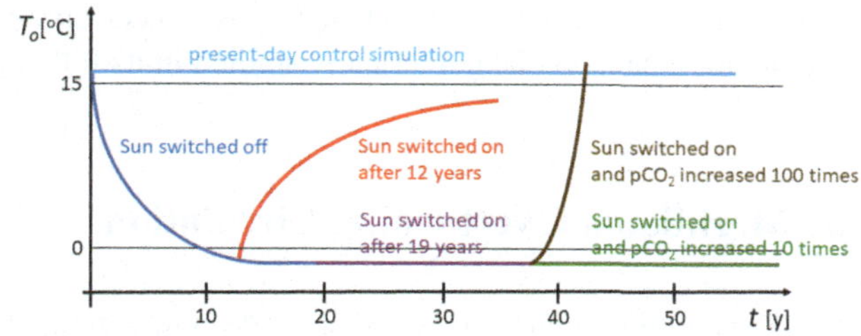


**Fig. 4.3** Sketch of time series of simulated global mean ocean temperature at 30 m depth. The light blue colour indicates the control simulation of present-day climate, dark blue: the solar constant is reduced to 0.1 per mil of its present-day value of $S_0 = 1365$ W/m$^2$ , magenta: after 19 years of strongly reduced solar constant, the solar constant is increased to 682.5 W/m$^2$ for the first year and to 1365 W/m$^2$ thereafter, red: the solar constant is set to its present-day value $S_0$ already after 12 years, brown: after 15 years of staying in a Snowball Earth state, the $CO_2$ concentration is set to 100 times of its present-day value, green: same as brown, but the $CO_2$ concentration is set to only ten times of its present-day value. This sketch is adapted from a figure in Marotzke and Botzet (2007). Please note that this figure is just a sketch of the original figure. The original figure also depicts the fluctuations of temperature

can cope with extreme boundary conditions. The authors tried to get the modelled Snowball Earth out of its frozen state by switching on the sun again. But, as already hypothesised by Wetherald and Manabe (1975), the Snowball state appeared to be a stable climate equilibrium in this MPI-ESM (see Fig. 4.3). Even a prescribed rise in atmospheric $CO_2$ by a factor of 10 from its present-day value did not melt the simulated Snowball. A rapid return to a green Earth within a few simulation years happened only when the $CO_2$ concentration was set to a level of 100 times the present-day value. This simulation clearly answered the question that also quasi-realistic climate models can produce multiple Snowball states and abrupt transitions between states.

The story continues. The study by Marotzke and Botzet (2007) triggered a number of follow-up studies with more elaborate versions of the MPI-ESM (e.g., Voigt et al. 2011; Voigt and Abbot 2012) and a newer version of model, the ICON-ESM (e.g., Ramme and Marotzke 2022). The study by Ramme and Marotzke (2022) is particularly noteworthy because, for the first time, a complete Snowball Earth cycle - although only with dynamic sea ice, but not dynamic ice sheets - was simulated. The study focuses on the effect of oceanic circulation and marine biogeochemistry on the transition from the Marinoan Snowball to the following super-greenhouse state. Thereby, the authors provided physically and chemically consistent constraints within which geological evidence has to be interpreted. Obviously, climate models are no longer only useful for investigating long-term equilibrium climate responses to prescribed perturbation as Hoffman et al. (2017) assumed, but are useful for studying millennial-scale and large-scale extreme climate transitions. And the story continues further. Some climate models of intermediate complexity (e.g., Calov et al. 2005, or of high complexity, e.g., Vizcaíno et al. 2008) reveal multiple climate states including dynamic inland-ice sheets. Hence, it is conceivable that realistic Snowball Earth dynamics can soon be numerically studied in its full complexity.

## 4.3 Atlantic Meridional Overturning Circulation

This section also starts with a prediction. Stommel (1961) investigated the circulation of water between two connected vessels. Each vessel is put into a larger vessel with constant, but different temperature **T**, −**T** and salinity **S**, −**S**, respectively (see Fig. 4.4). Through porous walls, temperature and salinity can diffuse into the water from the outer vessels into the water of the connected vessels. The temperature and salinity of the inner, connected vessels (Vessel 1 and Vessel 2) are labelled $T_1, T_2$ and $S_1, S_2$, respectively. The water in the inner vessels is stirred to maintain a uniform temperature and salinity in the respective vessels. Stommel focused on symmetric forms of the problem, hence he assumed that $T_1 = T, T_2 = -T$ and $S_1 = S, S_2 = -S$. Because of the temperature and salinity differences between the inner vessels, a water flux $q$ exists, which is governed by the conservation of heat and mass:

$$\frac{dT}{dt} = c\,(T - \mathbf{T})) - |2q|T \tag{4.4}$$

$$\frac{dS}{dt} = d\,(S - \mathbf{S})) - |2q|S \tag{4.5}$$

where $c$ and $d$ are the transfer coefficients of heat and salt, respectively.

This numerical experiment revealed three equilibrium states, two states (labeled **a** and **c** in Fig. 4.5) appear to be an attractor and the state (labeled **b**), a repellor.

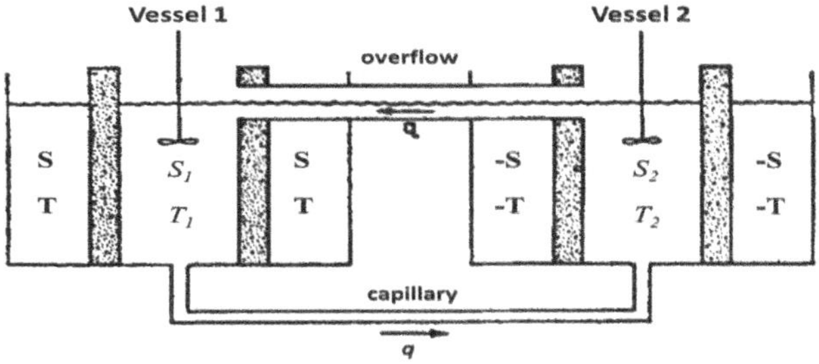


**Fig. 4.4** Set up of the two-vessel experiment by Stommel (1961). Two vessels containing a fluid with temperature and salinity $T_1$, $S_1$ and $T_2$, $S_2$, respectively are embedded in outer vessels with constant temperatures **T, S** and **−T, −S**. The inner and outer vessels are separated by porous walls. Because of the temperature and salinity differences between the inner vessels, a capillary flow $q$ is induced. The fluid in the inner vessels are stirred to maintain homogeneous temperature and salinity within each inner vessel. This figure is reprinted with modifications from Stommel (1961) with the permission of *Tellus A*

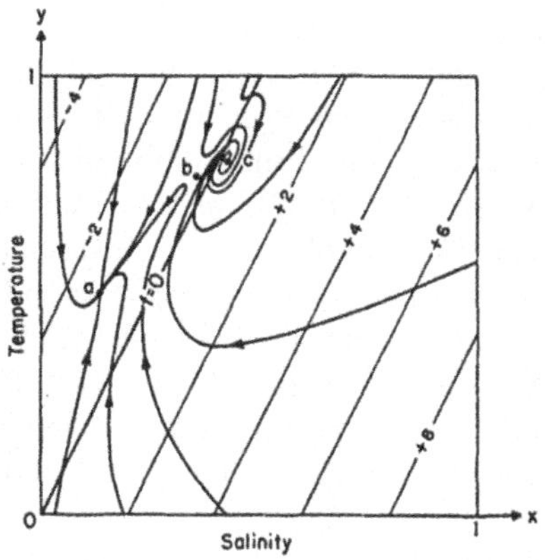


**Fig. 4.5** Phase diagram of the two-vessel problem presented by Stommel (1961). The y-axis indicates a non-dimensional temperature, and the x-axis, a non-dimensional salinity. A few integral curves are sketched to show the dynamics, specifically different equilibrium states, a stable fixed point **a**, an unstable (saddle) node **b**, and a stable spiral node **c**. This figure is taken from Stommel (1961) with the permission of *Tellus A*

Stommel (1961) noted that the mass flux $q$ occurs *sometimes in two different stable regimes, and may possibly be analogous to certain features of the oceanic circulation.*

Like in the case of the Snowball Earth, Stommel's prediction attracted a lot of interest by climate modellers when Hans Oeschger (Siegenthaler et al. 1984) and Willy Dansgaard (Dansgaard et al. 1982) detected rapid swings in temperature in Lake Gerzensee in Switzerland and in an ice core drilled in Greenland. More evidence of multiple rapid transitions between relatively warm phases (so-called interstadials) and relatively cold phases (so-called stadials) in Greenland during the last glacial cycle has been found (e.g., Grootes et al. 1993). The imprint of stadials and interstadials can be detected not only in Greenland ice cores, but in most palaeo-climate archives in the northern hemisphere and southern tropics, see Voelker et al. (2002). Because of its importance, the interstadials are now termed Dansgaard-Oeschger events. Some of the extreme cold events are associated with large calving events of ice bergs from the ice sheets in North America. The icebergs carried with them large amounts of debris, which were detected in marine sediments in the East Atlantic by Hartmut Heinrich . The discovery of the now-called "Heinrich events" is another significant contribution from a scientist in Hamburg (see Sect. 7.3).

When the first evidence of rapid temperature changes was seen in the Greenland ice cores, climate researchers suggested that these were somehow related to an instability of the thermohaline circulation in the Atlantic. *Does the ocean-atmosphere system have more than one stable mode of operation?* asked Wallace Broecker, Dorothy Peteet and David Rind (1985) which prompted a number of studies based on models of intermediate complexity (e.g., Marotzke et al. 1988).

The search for multiple states in the complex numerical models of global ocean circulation also started in the 1980s. Bryan (1986) used a model with idealised ocean basin geometry. The basin spanned a 60°-wide sector of the sphere extending from pole to pole, with a flat bottom at 5 km depth without sea ice. Frank Bryan found equilibrium thermohaline circulations symmetric or asymmetric to the equator, depending on an initial perturbation of the salinity distribution. Independently of each other, Manabe and Stouffer (1988) and Maier-Reimer and Mikolajewicz (1989) were the first to attack the problem with quasi-realistic circulation models. Manabe and Stouffer coupled models of atmospheric and oceanic circulation by using a flux correction to prevent the coupled model from drifting into unrealistic states. The results of their simulations reveal two clear stable equilibria, one with a strong AMOC (Atlantic Meridional Overturning Circulation) and one without a large-scale AMOC, but just a very small AMOC within the northern tropical Atlantic.

Maier-Reimer and Mikolajewicz (1989) (for a more comprehensive discussion see Maier-Reimer et al. 1993) tested the sensitivity of their complex ocean circulation model to small perturbations in fresh water fluxes at different locations of the North American Atlantic coast. Depending on the location and the amount of fresh water put into the North Atlantic, they found either a marginal response or a rapid transition to a new state, as seen in the North Atlantic meridional heat flux (see Fig. 4.6). It is worth noting that the ocean model by Maier-Reimer and Mikolajewicz (1989) appears to reveal a realistic response of the AMOC to freshwater perturbations when comparing their experiments with estimates from palaeo-climate reconstructions of the so-called Younger Dryas. The latter event was a cold spell between approximately 12,900 and 11,700 years ago before the onset of the current warm phase, the Holocene. The proposition by Valdes (2011) that complex, quasi-realistic models are constructed for stability and respond too

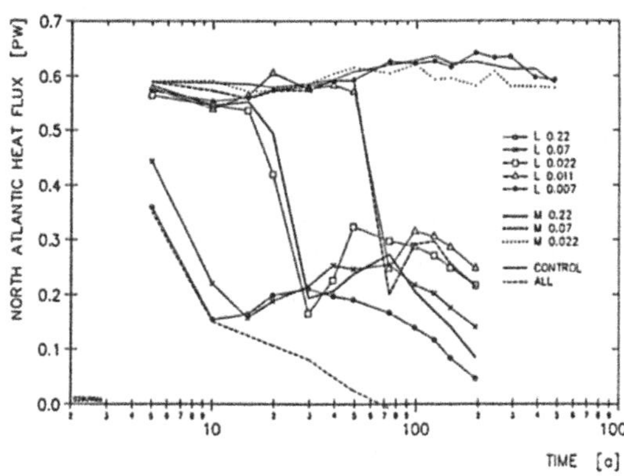


**Fig. 4.6** Simulated time series of the North Atlantic heat flux in water hosing experiments. *L* and *M* denote release of freshwater at St. Lawrence River and Mississippi, respectively. The numbers indicate the strength of the fresh water inflow in Sverdrup $\left[10^6\ m^3/s\right]$. "All" denotes the experiment with six times the mean deglaciation rate. This figure is taken from Maier-Reimer and Mikolajewicz (1989) (Copyright Max-Planck-Institut für Meteorologie, Hamburg)

insensitively to realistic perturbation of the system does not seem to hold for the ocean model of Maier-Reimer and Mikolajewicz (1989).

The story continues. More studies using quasi-realistic models tackle the issue of rapid reorganizations of the AMOC (e.g., Liu et al. 2009; Kapsch et al. 2022). Some of the models even incorporate ice-sheet dynamics to study the interaction between ice sheets, ocean circulation and the climate during Heinrich events, for example Ziemen et al. (2019).

## 4.4 Climate-Vegetation Interaction

The third example begins with a prediction, which is somewhat hidden. Watson and Lovelock (1983) aimed at demonstrating that the biosphere can interact with its environment *by and for the biosphere*. The biosphere would not only be affected by its environment, but would also be able to modify its environment, thereby creating conditions optimally for its growth. Even under changing external boundary conditions, the interaction between the biosphere and the environment would produce more or less homeostatic conditions. For example, in a warming climate, the biosphere would tend to cool its environment to keep optimum conditions for its growth, and *vice versa*, in a cooling climate, the biosphere would try to warm the environment. Hence Watson and Lovelock's paper focussed on explaining homeostasis, pretty much the opposite of abrupt changes and multiple equilibria.

To demonstrate that climate-biosphere homeostasis is not related to some teleological foresight of the biosphere but emerges as a consequence of stabilizing physical feedback processes, Watson and Lovelock (1983) set up a simple climate-vegetation model. The model combines a climate model and a dynamic vegetation model. Watson and Lovelock called their model "Daisyworld" model to emphasize that it was meant to only demonstrate the effect of stabilizing feedbacks in a highly idealized world. The model was not meant as any realistic description of our Earth system. The climate of the Daisyworld was computed in the same way as the Earth's energy balance Eq. (4.2):

$$0 = Q_p(1 - \alpha_p) - I(T_p) \tag{4.6}$$

where $Q_p$ is the net incoming shortwave radiation and $T_p$ is its temperature. Daisyworld's planetary albedo $\alpha_p$ and outgoing heat radiation $I(T_p)$ are assumed to linearly depend on local albedo values and local temperatures:

$$\alpha_p = \sum_i V_i \alpha_i$$

$$T_p = \sum_i V_i T_i$$

where $V_i$ are the fractions $V_i, i = 0, 1, 2, ...$ of the Daisyworld planet covered by either bare ground ($i = 0$) or vegetation of a type ($i = 1, 2, ..$). The original Daisyworld model used one or two different plant types, *daises* with an albedo $\alpha_1$ lower than the albedo $\alpha_0$ of the bare ground and *daises* with an albedo $\alpha_2$ higher than $\alpha_0$.

The vegetation part of Watson and Lovelock's was the model of competing species developed in 1926 by Vito Volterra, based on a model of autocatalytic chemical reactions proposed by Alfred Lotka in 1911. The so-called Lotka–Volterra equations for Daisyworld read:

$$\frac{dV_i}{dt} = V_i \left[ \left( 1 - \sum_{j=1}^{N} V_i \right) \beta(T_i) - \gamma \right] \tag{4.7}$$

where $\beta(T_i)$ is a growth function which possesses a maximum value, i.e., a temperature of optimum growth, and which is defined as a positive function between two temperature values within which plants can exist and being zero otherwise. $\gamma$ is a constant mortality rate.

The Lotka-Volterra equations are known to yield multiple equilibria, or fixed points, some are stable, others are unstable. Hence Watson and Lovelock's model should reveal abrupt transitions emerging from the loss of the stability of one of the equilibrium states. This behaviour was explored by Weber (2001), for example. The bifurcation diagramme is depicted in Fig. 4.7.

The red straight line in Fig. 4.7 depicts the planetary temperature $T_p$ of an uninhabited Daisyworld. The temperature $T_p$ of the uninhabited planet rises monotonously with insolation $L$. However, in the range between $L_2$ and $L_5$ when

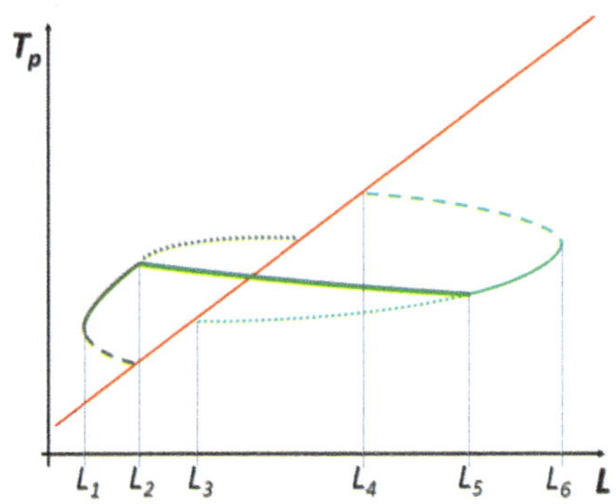


**Fig. 4.7** Stable and unstable equilibrium solutions of the vegetation cover in the Daisyworld model analysed by Weber (2001). The equilibrium temperature as a function of the solar luminosity: no daisies (red line), low- and high-albedo daisies (thick green line), low-albedo daisies only (dotted and dashed dark-green line) and high-albedo daisies only (dotted and dashed light-green line). Stable states are indicated by full lines and dotted lines. This figure is redrawn with modifications from Weber (2001)

both, low-albedo plants and high-albedo plants can exist, then $T_p$ slightly changes with increasing insolation (thick green line); it even slightly decreases. This is the homeostasis proposed and described by Watson and Lovelock (1983).

More interesting from the dynamical point of view are the dashed and dotted lines, which show the non-trivial unstable and stable equilibria $V_i \neq 0$ of low-abedo and high-albedo daisies, respectively. At low insolation $L < L_1$, no daisies can exist because the climate is just too cold for any daisies to grow. If insolation increases beyond $L_1$, then three equilibria for low-albedo daisies exist: the trivial state $V_1 = 0$ with $T_1 = T_p$ of the uninhabited plant (red line), and two non-trivial states $V_1 \neq 0$ with $T_1 > T_p$. One of the non-trivial solutions is stable (dark green curve), and the other is unstable to infinitesimally small perturbations (dashed dark-green curve). The latter acts as a repellor between the two stable states. Therefore, low-albedo daisies emerge abruptly at an insolation $L = L_1$, if for some reason, an abundant seed of low-albedo daisies is available. Alternatively, if there are no abundant seeds, then an abrupt emergence of low-albedo daisies only occurs at a higher level of insolation $L = L_2$. At higher insolation values between $L_2$ and $L_5$, low- and high-albedo can grow on Daisyworld, until above insolation values of $L_5$, the climate gets too hot for low-albedo daisies, and only high-albedo daisies remain, because they better reflect radiation and, hence, cool the local climate. If only low-albedo daisies would be allowed to grow, then the temperature would follow the dotted dark-green curve. In turn, if only high-albedo daisies were available in the Daisyworld, then these high-albedo daisies would appear gradually at insolation values $L = L_3$ (dotted light-green curve). Above $L_5$, three solutions emerge for high-albedo daisies, until the insolation becomes too strong ($L > L_6$), and all daisies vanish.

If the Daisyworld starts with strong insolation $L > L_6$ and if the insolation decreases, then at an insolation $L < L_6$, three solutions of the Daisyworld vegetation appear with the possibility of abrupt colonisation by high-albedo daisies. And the dynamics evolve in the opposite direction as explained above.

Hence, daisies can appear abruptly when life on Daisyworld starts, and at the end of the lifespan of Daisyworld's biosphere, daisies vanish abruptly. The Daisyworld model suggests that there is not only a homeostasis, but abrupt climate and vegetation changes may also appear, just because of the interaction by and for the biosphere.

Would this *parable of Daisyworld* (Watson and Lovelock 1983) also hold for more complex climate-vegetation models describing the real world? In principle, it does. Independently of the Daisyworld discussion, Claussen (1997) found multiple climate-vegetation equilibria when coupling the quasi-realistic model of atmospheric circulation ECHAM (Deutsches Klimarechenzentrum et al. 1992) with a diagnostic vegetation model, the model BIOME 1.0 (Prentice et al. 1992), When the climate-biome simulations were initialized with continents covered with present-day biomes or continents with deserts, then a solution appeared that was very close to the present-day climate and biome distribution. However, initialization with all continents covered by present-day vegetation, but with tropical forest in the Sahara and Sahel, led a solution with a much greener Sahara than today mainly covered

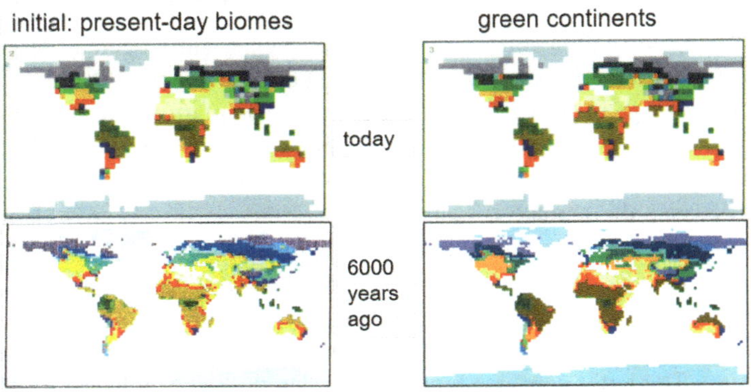


**Fig. 4.8** Simulated multiple climate-vegetation states in present-day climate (upper row) and a "green" climate-vegetation state in mid-Holocene climate, some 6000 years ago (lower row). The simulations were undertaken by Claussen (1997) for present-day climate and Claussen and Gayler (1997) for mid-Holocene climate by using the same model of atmosphere circulation coupled to the BIOME-1 model of Prentice et al. (1992). The figures on the left-hand side are equilibrium states when initializing the simulations with present-day biome cover, and on the right-hand side, with a tropical forest biome cover in the Sahara and Sahel. These figures are taken from Claussen (1997) and Claussen and Gayler (1997) with the permission of *Springer Nature*

with some xerophytic shrubs and savanna (Fig. 4.8 upper row, right figure). Further experimentation with different initial vegetation cover on all continents showed that multiple equilibria only emerged in the Sahara/Sahel region (Claussen 1998). The occurrence of multiple equilibria in that region can be attributed to a large part to a perturbation in the vertical radiation budget of the subtropics caused by the large difference in albedo values between bright deserts and rather dark vegetation, see Charney (1975) and Claussen (1997).

After Claussen (1997) reported about multiple climate-vegetation states, others followed. Wang and Eltahir (2000) coupled a zonally-symmetric model of the tropical atmosphere with a dynamic vegetation model and found bistability. Likewise Oyama and Nobre (2003) detected multiple equilibria of tropical forest versus caatinga, a xeric shubland and thorn forest, and savanna in Amazonia.

Kleidon et al. (2007) criticised the emergence of multiple climate-biome equilibria because the use of discretised vegetation or biomes necessarily leads to a set of discrete solutions without any continuous spectrum of solutions in between. This proposition is certainly valid. However, multiple climate-vegetation equilibria appear to be possible when using a model of continuous vegetation—as will be shown below.

The story of multiple climate-vegetation equilibria became even more interesting when Claussen and Gayler (1997) found only one solution with a *green Sahara* independent of initial conditions, when the climate-biome model was run for insolation conditions that prevailed some 6000 years ago (Fig. 4.8 lower row). During the mid-Holocene between some 8200 and 4200 years ago, the perihelion of the Earth's orbit and the tilt of the Earth's axis were different from its present-day values, which led to a stronger insolation on the Northern Hemisphere during the boreal summer months and a weaker insolation, during the boreal winter. The difference in the manifold of equilibrium solution as a function of insolation suggests the existence of a threshold in orbital forcing and the possibility of an abrupt transition between a *green Sahara* and a Sahara as it exists today, as demonstrated by Brovkin et al. (1998).

Brovkin et al. (1998) set up a simple dynamical system of climate-vegetation interaction in arid and semi-arid regions. The model assumes that vegetation tries to adjust to an equilibrium vegetation cover $V^*$ within a time scale $\tau$. $V^*$ depends on precipitation $P$ only, and, in turn, precipitation changes with vegetation as is the case in the Sahara/Sahel region:

$$\frac{dV}{dt} = \frac{V - V^*(P)}{\tau} \tag{4.8}$$

$$P^*(V) = P_0 + bV(t) \tag{4.9}$$

where $V^*$ is the vegetation cover in equilibrium with precipitation. The function $V^* = V^*(P)$ is steady and monotonous. $P_0$ denotes a large-scale background precipitation which can depend on some external parameters, but not on the regional climate.

By fitting the parameters of the simple climate-vegetation model to the results of the quasi-realistic model for present-day (Claussen 1997) and mid-Holocene (Claussen and Gayler 1997) conditions, Brovkin et al. (1998) found the bifurcation diagram depicted in Fig. 4.9, and concluded that an abrupt transition from the mid-Holocene *green Sahara* to the present-day Sahara should have occurred between 6000 to 4000 years ago.

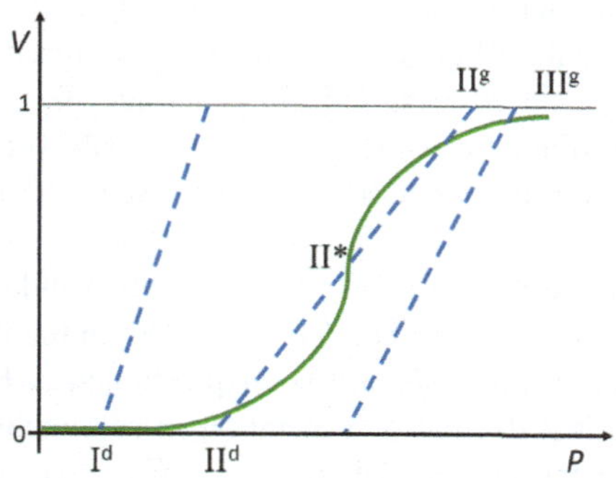


**Fig. 4.9** Schematic draft of the model by Brovkin et al. (1998). The solid curve represents the dependence of equilibrium vegetation $V*$ on annual mean precipitation $P$. The dashed line represents the linear model of precipitation $P^*$ as a function of vegetation cover $V$. Cases I and III correspond to equilibrium solutions: desert $I^d$ and green $III^g$, respectively; cases II correspond to the stable desert $II^d$ and the green $II^g$ equilibrium, and $II^*$, to the unstable solution. This figure is redrawn with modifications from Brovkin et al. (1998)

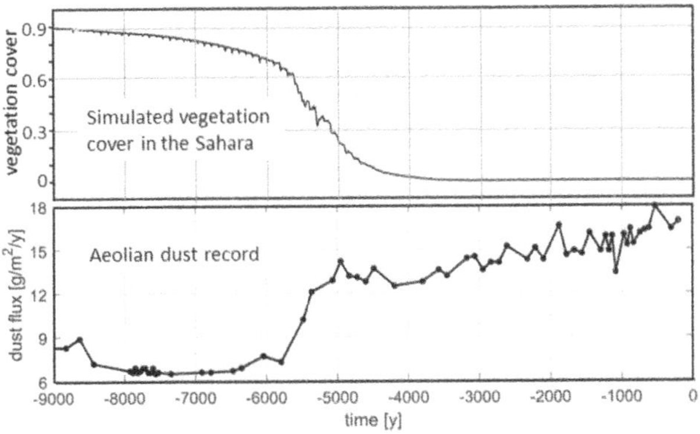


**Fig. 4.10** Fractional vegetation coverage simulated by using the model of intermediate complexity the CLIMBER-2 (Claussen et al. 1999) (upper figure) and the Site 658C marine record of West African eolian dust supply variations (lower figure). Please note that the scale of the lower figure is inverted to highlight the similarity between predicted vegetation changes and reconstructed dust changes. This figure is redrawn with modifications from Claussen et al. (1999), Aeolian dust data are taken from deMenocal et al. (2000)

Using the CLIMBER-2 model, a global model of intermediate complexity coupling dynamics of the atmosphere, the ocean and the vegetation at a geographically coarse resolution, Claussen et al. (1999) also found multiple climate-vegetation states in the Sahara/Sahel region and a rapid transition towards the present-day desert conditions around 5500 years ago. The CLIMBER-2 model was designed as a statistical dynamical model in which weather variability does not exist. Therefore, perturbations in simulated vegetation cover were artificially imposed on the model to demonstrate that the emergence of multiple equilibria in CLIMBER-2 was not just the result of missing climate noise.

A year after the prediction of a rapid expansion of the Sahara in the mid-Holocene, deMenocal et al. (2000) detected an abrupt increase in dust concentration within only 300 years, centred around 5500 years ago, in marine sediment cores near the West African Atlantic coast (Fig. 4.10). This palaeo-climatic evidence was considered as validation of the prediction of a collapse of the African humid period (deMenocal et al. 2000).

But, fortunately, the story continues. The seeming validation of a collapse of the African humid period and the *green Sahara* based on the prediction of multiple climate-vegetation equilibrium states was soon questioned by Kröpelin et al. (2008). They saw a rather gradual decline of vegetation in sediments of the Lake Yoa in the Ounianga Kébir in the central Sahara. This, of course, was also not the end of the story. By analysing all available palaeo data in the northern African region, Shanahan et al. (2015) found a time-transgressive termination of the African humid period with an end earlier in the north than in the south and earlier in the east than in the west. Rapid vegetation changes were found only in a few locations. Subsequent studies using the quasi-realistic MPI-ESM could not only provide an explanation for the time-transgressive transition of the mid-Holocene Sahara (Dallmeyer et al. 2020), but could also demonstrate the link between the abrupt increase in the dust flux and the rapid changes in the landscape in some parts of the western Sahara (Egerer et al. 2018). Obviously, a *collapse* of the entire green Sahara (Brierley et al. 2018) happens only in simple models or in models with a poor geographical resolution like in CLIMBER-2 in which the Sahara is represented by only one grid box.

This is not the end of the story either. Most complex Earth system models use vegetation models with a rather coarse classification of different plant types. Thereby, the natural plant-functional diversity is grossly underestimated. In some regions like the mid-Holocene Sahara, there is no modern analogue of plant societies (Watrin et al. 2009). Hence, the vegetation in such regions can be described by current models only to some limited degree. And even more important, plant functional diversity has a strong impact on climate-vegetation interaction. In models with high plant functional diversity, the occurrence of multiple equilibria may vanish (Claussen et al. 2013).

## 4.5 Concluding Remarks

Multiple climate equilibria, abrupt transitions, climate catastrophes and tipping points have fascinated climate researchers ever since they detected them first in theory and models of the 1960s and later in palaeo-climate reconstructions. Klaus Hasselmann was sceptical about the *unstable feedback models* that presumably *tend to predict climatic variations 'as flip-flop transitions' and therefore fail to reproduce the observed continuous spectrum of climatic variability* (Hasselmann 1976). The exploration of quasi-realistic models of the Earth's climate system basically refuted Hasselmann's sceptical remark made in 1976. The occurrence of multiple states in relation to Snowball Earths was confirmed (Sect. 4.2), which led to a new era of complex Snowball Earth modelling, thereby providing more realistic, physical and geochemically consistent constraints for the interpretation of geological evidence. Ernst Maier-Reimer was the first to simulate rapid transitions in the AMOC triggered by small perturbations in freshwater flux (Sect. 4.3). And the detection of multiple equilibrium states in the complex climate-vegetation models (Sect. 4.4) led to a new view on climate-biosphere dynamics. Of course, the story continues. Even quasi-realistic, so-called state-of-the-art models are models of a *high degree of idealization* in comparison with the many more degrees of freedom the natural climate system reveals. Hence global climate and Earth system model run at a very high resolution of a few kilometres are pursued (e.g. Hohenegger et al. 2023). Such *digital twins* of the natural Earth's climate system will likely provide new insight into fine-scale processes which are relevant at the global scale (e.g. Vogel et al. 2022). Whether or not these models will change the picture of multiple equilibria and rapid transitions in the Earth's climate system remains to be seen.

## References


Brierley C, Manning K, Maslin M (2018) Pastoralism may have delayed the end of the green Sahara. Nat Commun 9(1):4018

Broecker WS, Peteet DM, Rind D (1985) Does the ocean–atmosphere system have more than one stable mode of operation? Nature 315(6014):21–26

Brovkin V, Claussen M, Petoukhov V, Ganopolski A (1998) On the stability of the atmosphere-vegetation system in the Sahara/Sahel region. J Geophys Res: Atmos 103(D24):31613–31624

Bryan F (1986) High-latitude salinity effects and interhemispheric thermohaline circulations. Nature 323(6086):301–304

Budyko MI (1969) The effect of solar radiation variations on the climate of the earth. Tellus 21(5):611–619

Calov R, Ganopolski A, Claussen M, Petoukhov V, Greve R (2005) Transient simulation of the last glacial inception. Part I: glacial inception as a bifurcation in the climate system. Clim Dyn 24:545–561

Charney JG (1975) Dynamics of deserts and drought in the Sahel. Q J R Meteorol Soc 101(428):193–202

Claussen M (1997) Modeling bio-geophysical feedback in the African and Indian monsoon region. Clim Dyn 13:247–257

Claussen M (1998) On multiple solutions of the atmosphere–vegetation system in present-day climate. Glob Change Biol 4(5):549–559

Claussen M, Gayler V (1997) The greening of the Sahara during the mid-Holocene: results of an interactive atmosphere-biome model. Glob Ecol Biogeogr Lett 6:369–377

Claussen M, Kubatzki C, Brovkin V, Ganopolski A, Hoelzmann P, Pachur HJ (1999) Simulation of an abrupt change in Saharan vegetation in the mid-Holocene. Geophys Res Lett 26(14):2037–2040

Claussen M, Bathiany S, Brovkin V, Kleinen T (2013) Simulated climate–vegetation interaction in semi-arid regions affected by plant diversity. Nat Geosci 6(11):954–958

Dallmeyer A, Claussen M, Lorenz SJ, Shanahan T (2020) The end of the African humid period as seen by a transient comprehensive Earth system model simulation of the last 8000 years. Clim Past 16(1):117–140

Dansgaard W, Clausen HB, Gundestrup N, Hammer CU, Johnsen SF, Kristinsdottir P, Reeh N (1982) A new Greenland deep ice core. Science 218(4579):1273–1277

deMenocal P, Ortiz J, Guilderson T, Adkins J, Sarnthein M, Baker L, Yarusinsky M (2000) Abrupt onset and termination of the African humid period: rapid climate responses to gradual insolation forcing. Quat Sci Rev 19(1–5):347–361

Deutsches Klimarechenzentrum, et al (1992) The ECHAM3 atmospheric general circulation model. Techn Rep 6

Donnadieu Y, Ramstein G, Fluteau F, Roche D, Ganopolski A (2004) The impact of atmospheric and oceanic heat transports on the sea-ice-albedo instability during the Neoproterozoic. Clim Dyn 22:293–306

Egerer S, Claussen M, Reick C (2018) Rapid increase in simulated North Atlantic dust deposition due to fast change of northwest African landscape during the Holocene. Clim Past 14(7):1051–1066

Fraedrich K (1979) Catastrophes and resilience of a zero-dimensional climate system with ice-albedo and greenhouse feedback. Q J R Meteorol Soc 105(443):147–167

Grassl H (1981) The climate at maximum entropy production by meridional atmospheric and oceanic heat fluxes. Q J R Meteorol Soc 107(451):153–166

Grootes PM, Stuiver M, White J, Johnsen S, Jouzel J (1993) Comparison of oxygen isotope records from the GISP2 and GRIP Greenland ice cores. Nature 366(6455):552–554

Hasselmann K (1976) Stochastic climate models. Part I. Theory. Tellus 28:473–485

Hoffman PF, Schrag DP (2002) The snowball Earth hypothesis: testing the limits of global change. Terra Nova 14(3):129–155

Hoffman PF, Abbot DS, Ashkenazy Y, Benn DI, Brocks JJ, Cohen PA, Cox GM, Creveling JR, Donnadieu Y, Erwin DH, et al (2017) Snowball Earth climate dynamics and Cryogenian geology-geobiology. Sci Adv 3(11):e1600983

Hohenegger C, Korn P, Linardakis L, Redler R, Schnur R, Adamidis P, Bao J, Bastin S, Behravesh M, Bergemann M, et al (2023) ICON-sapphire: simulating the components of the Earth system and their interactions at kilometer and subkilometer scales. Geosci Model Dev 16(2):779–811

Kapsch ML, Mikolajewicz U, Ziemen F, Schannwell C (2022) Ocean response in transient simulations of the last deglaciation dominated by underlying ice-sheet reconstruction and method of meltwater distribution. Geophys Res Lett 49(3):e2021GL096767

Kirschvink JL (1992) Late Proterozoic low-latitude global glaciation: the snowball Earth. In: Schopf J, Klein C, Maris DD (eds) The proterozoic biosphere: a multidisciplinary study. Cambridge University Press, Cambridge, pp 51–52

Kleidon A, Fraedrich K, Low C (2007) Multiple steady-states in the terrestrial atmosphere-biosphere system: a result of a discrete vegetation classification? Biogeosciences 4(5):707–714

Kröpelin S, Verschuren D, Lézine AM, Eggermont H, Cocquyt C, Francus P, Cazet JP, Fagot M, Rumes B, Russell JM, et al (2008) Climate-driven ecosystem succession in the Sahara: the past 6000 years. Science 320(5877):765–768

Lemke P (1977) Stochastic climate models, Part 3. Application to zonally averaged energy models. Tellus 29:385–392. https://doi.org/10.3402/tellusa.v29i5.11371

Liu Z, Otto-Bliesner B, He F, Brady E, Tomas R, Clark P, Carlson A, Lynch-Stieglitz J, Curry W, Brook E, et al (2009) Transient simulation of last deglaciation with a new mechanism for Bølling-Allerød warming. Science 325(5938):310–314

Lucarini V, Fraedrich K, Lunkeit F (2010) Thermodynamic analysis of snowball earth hysteresis experiment: efficiency, entropy production and irreversibility. Q J R Meteorol Soc A J Atmos Sci Appl Meteorol Phys Oceanogr 136(646):2–11

Maier-Reimer E, Mikolajewicz U (1989) Experiments with an OGCM on the cause of the Younger Dryas. Max-Planck-Institut für Meteorologie Report 39

Maier-Reimer E, Mikolajewicz U, Hasselmann K (1993) Mean circulation of the hamburg LSG OGCM and its sensitivity to the thermohaline surface forcing. J Phys Oceanogr 23(4):731–757

Manabe S, Stouffer RJ (1988) Two stable equilibria of a coupled ocean-atmosphere model. J Clim 1(9):841–866

Marotzke J, Botzet M (2007) Present-day and ice-covered equilibrium states in a comprehensive climate model. Geophys Res Lett 34(16)

Marotzke J, Welander P, Willebrand J (1988) Instability and multiple steady states in a meridional-plane model of the thermohaline circulation. Tellus A Dyn Meteorol Oceanogr 40(2):162–172

Oyama MD, Nobre CA (2003) A new climate-vegetation equilibrium state for tropical South America. Geophys Res Lett 30(23)

Paltridge GW (1975) Global dynamics and climate-a system of minimum entropy exchange. Q J R Meteorol Soc 101(429):475–484

Prentice IC, Cramer W, Harrison SP, Leemans R, Monserud RA, Solomon AM (1992) A global biome model based on plant physiology and dominance, soil properties and climate. J Biogeogr 19:117–134

Ramme L, Marotzke J (2022) Climate and ocean circulation in the aftermath of a Marinoan snowball Earth. Clim Past 18(4):759–774

Sellers WD (1969) A global climatic model based on the energy balance of the earth-atmosphere system. J Appl Meteorol Climatol 8(3):392–400

Shanahan TM, McKay NP, Hughen KA, Overpeck JT, Otto-Bliesner B, Heil CW, King J, Scholz CA, Peck J (2015) The time-transgressive termination of the African Humid Period. Nat Geosci 8(2):140–144

Siegenthaler U, Eicher U, Oeschger H, Dansgaard W (1984) Lake sediments as continental $\delta^{18}$O records from the glacial/post-glacial transition. Ann Glaciol 5:149–152

Stommel H (1961) Thermohaline convection with two stable regimes of flow. Tellus 13(2):224–230

Stone P, Yao M (2004) The ice-covered earth instability in a model of intermediate complexity. Clim Dyn 22(8):815–822

Valdes P (2011) Built for stability. Nat Geosci 4(7):414–416

Van Nes EH, Arani BM, Staal A, van der Bolt B, Flores BM, Bathiany S, Scheffer M (2016) What do you mean, 'tipping point'? Trends Ecol Evol 31(12):902–904

Vizcaíno M, Mikolajewicz U, Gröger M, Maier-Reimer E, Schurgers G, Winguth AM (2008) Long-term ice sheet–climate interactions under anthropogenic greenhouse forcing simulated with a complex earth system model. Clim Dyn 31:665–690

Voelker AH, et al (2002) Global distribution of centennial-scale records for marine isotope stage (MIS) 3: a database. Quat Sci Rev 21(10):1185–1212

Vogel R, Albright AL, Vial J, George G, Stevens B, Bony S (2022) Strong cloud–circulation coupling explains weak trade cumulus feedback. Nature 612:1–5

Voigt A, Abbot D (2012) Sea-ice dynamics strongly promote Snowball Earth initiation and destabilize tropical sea-ice margins. Clim Past 8(6):2079–2092

Voigt A, Abbot DS, Pierrehumbert RT, Marotzke J (2011) Initiation of a Marinoan Snowball Earth in a state-of-the-art atmosphere-ocean general circulation model. Clim Past 7(1):249–263

Wang G, Eltahir EA (2000) Biosphere—atmosphere interactions over West Africa. II: multiple climate equilibria. Q J R Meteorol Soc 126(565):1261–1280

Watrin J, Lézine A, Hély C, et al (2009) Plant migration and ecosystems at the time of the "green Sahara". CR Geosci 341:656–670

Watson AJ, Lovelock JE (1983) Biological homeostasis of the global environment: the parable of daisyworld. Tellus B Chem Phys Meteorol 35(4):284–289

Weber S (2001) On homeostasis in daisyworld. Clim Change 48(2–3):465–485

Wetherald RT, Manabe S (1975) The effects of changing the solar constant on the climate of a general circulation model. J Atmos Sci 32(11):2044–2059

Ziemen FA, Kapsch ML, Klockmann M, Mikolajewicz U (2019) Heinrich events show two-stage climate response in transient glacial simulations. Clim Past 15(1):153–168

# Chapter 5
# Contributions to Modelling Techniques

**Robert Sausen, Eduardo Zorita, Hans von Storch,
Martin Claussen, and Martin Heimann**

**Abstract** Reliably simulating climate by means of comprehensive models has required dedicated numerical techniques. A few of them were developed in Hamburg. (1) Walter Hansen introduced numerical modelling in oceanography, which was one of the foundations of climate modelling in Hamburg. (2) Ernst Maier-Reimer's Large Scale Geostrophic model of the ocean enabled millennial simulations, e.g., for studying the oceanic carbon cycle. (3) The flux correction (anomaly coupling) method allowed decadal simulations of coupled atmosphere-ocean models without a model drift and, hence, more reliable estimates of global warming.

## 5.1 Walter Hansen—Replacing Physical Models by Numerical Models[1]

Walter Hansen (1909–1991; see also Fig. 5.1) was the first professor at the Institut für Meereskunde (Institute of Oceanography) at the University of Hamburg from 1956 until 1977.[2] His main achievement was the design of "hydrodynamical-numerical" modelling (HN modelling) in oceanography, which is said to have earned him the title "nestor of HN modelling" by Kirk Bryan. This approach substituted the use of mechanical models, which had dominated engineering applications.

In his commemorative publication on the occasion of Hansen's 80th birthday Sündermann (1989) stated "*Hansen recognised with almost prophetic foresight the possibilities that lie in the mathematical modelling of physical processes in the sea. ... It was and is Hansen's lasting merit that he believed in the meaningfulness also of the numerically manipulated equations of determination and next took on the immense manual computational work. His teacher ..., also a respected oceanographer, advised him at that time against a hopeless undertaking: Already*

[1] This Section obtained substantial input from Jürgen Sündermann.

[2] https://www.hpk.uni-hamburg.de/resolve/id/cph_person_00000912.

 
H. von Storch et al., *Climate Science Concepts Born in Hamburg*,
https://doi.org/10.1007/978-3-031-81108-1_5

**Fig. 5.1** One of the rare photos of Walter Hansen, provided by Jürgen Sündermann

*the differences of two calculated water levels or velocities would be extremely inaccurate, but the differences of the differences would have no real meaning at all. Hansen did not allow his missionary urge to be interfered with. He calculated tidal waves in the Ems for the first time still in the forties, in 1952 the tides of the North Sea and already before (1949) those of the North Atlantic. The success proved him right. Truly, the vastness of the ocean could be forced into model equations, the solution of which yielded a new wealth of information in the interior of the seas that could not be obtained by measurements alone and certainly not with a comparable expenditure of material. The ability of the models to perform sensitivity analyses (how does the ocean react to ... ?) and prognostic scenarios (what happens if ... ?) proved to be particularly advantageous and gradually indispensable.*"

The worldwide acceptance of the hydrodynamic-numerical (HN) modelling became manifest at an international symposium in Hamburg in September 1961 (Institut für Meereskunde 1962), with contributions of, among others, G.W. Platzmann, P. Welander, J.J. Dronkers, A. Robinson, or H. Friedrichs. Even if not included in the proceedings, Jürgen Sündermann remembers that other illustrious oceanographers, such as W. Munk and M. Abbott, were present.

**Fig. 5.2** Günter Fischer in his institute at the Universität Hamburg. Photo provided by Hans von Storch

Sündermann continues: "*In the last years of his active career, Hansen was able to observe with satisfaction the worldwide triumph of his idea. Model calculations ... had become an integral part of marine research, they were and are equally used in global circulation studies, for the investigation of the North Sea ecosystem, for small-scale studies in estuaries, rivers and lakes. This trend has been aided, of course, by the explosive development of electronic computing ... Hansen had the intuition to make new methods ... available just as the first computers were looking for applications. He rushed to Stockholm in 1956, where one of the first European large-scale computing facilities was located.*" Using these facilities, he published (Hansen 1956b) examples of "*a method* (that) *is developed for determining the movements in such areas; morever, the results of a numerical computation of an especially heavy storm surge ... and of the tides in an estuary ... are discussed. By combining these results with numerical weather prediction it is, in principle, possible to carry out numerical prediction of sea level.*"—the HN method.

Interestingly, the already mentioned Günter Fischer (Fig. 5.2) used also the computer resources in Stockholm and published in 1959 an article on modelling tidal waves and currents in the North Sea, with a special focus on the numerical methodology (Fischer 1959). This activity by Günter Fischer was an early effort which much later led to building the climate model ECHAM in Hamburg.

**Fig. 5.3** Hydraulical model of the Jade Bay (Sündermann and Vollmers 1972)

To comprehend the importance of Hansen's Ansatz, one must take into account the prevalent practices in oceanography and hydraulic engineering during that time, which relied heavily on physical models, i.e., scaled-down versions of the actual conditions (see e.g., Fig. 5.3). Creating such models was a laborious and expensive task, unlike numerical models, which were comparatively more cost-effective and allowed researchers to study phenomena that were not possible with hydraulic models, such as the rotation of the earth.

When the numerical model was run without the Coriolis force, it produced a completely symmetric result that was consistent with the hydraulic model. However, when the Coriolis force was activated in the numerical model, the symmetry broke down, and a distinct current pattern emerged (Sündermann and Vollmers 1972; Müller and von Storch 2004), demonstrating Hansen's original idea that numerical models were superior to physical models.

During that era, publishing papers in English, particularly in oceanography and hydraulic engineering, was not a common practice in Germany. The first known paper by Hansen was published in 1940, although it was unrelated to HN-modelling, which he eventually documented in Hansen (1956a).

## 5.2 An Ocean General Circulation Model for Long-Term Climate Studies: LSG by EMR[3,4]

The finding from ice-core measurements that the atmospheric $CO_2$ concentration was about one third lower 20,000 years ago compared to the Holocene spurred the scientific quest to identify possible mechanisms (see Chap. 3). Obviously, changes in global ocean circulation could have been responsible. A global ocean model that could simulate the circulation over thousands of years was needed to explore long-term climate variations, such as glacial cycles and to explain the natural positive feedback of the carbon cycle on the climate system on such time scales.

At that time, i.e., in the 1980s, global ocean models usually were based on the full set of equations of motion (Navier-Stokes equations, or so-called "primitive" equations) and could at best, even with a coarse horizontal resolution of several degrees in latitude and longitude, be integrated for a few hundred years only. However, the turnover time of the ocean amounts to about 1000–1500 years. Model runs over only a few centuries would not be in steady-state, i.e., the simulated state variables such as temperature and salinity would still drift and no sensitivity experiments to a stable baseline could be carried out. Further, due to the limited computational capacity in the late 1980s, immense costs and waiting times were associated with numerically integrating global ocean models. Hence, model runs could not be easily repeated to adjust model parameters and other tuning tasks. Further, a keen interest existed in analysing and possibly predicting slow natural climate fluctuations forced by random noise in driving factors of the ocean (cf. Hasselmann 1976; Sect. 2.3.2) and possibly constrained by oceanic system properties such as the shape of ocean basins and associated planetary wave characteristics.

Was there someone in the then still quite limited but exquisite staff at the Max Planck Institute of Meteorology at Hamburg that could carry out the as then considered "impossible task" of developing a computationally economical global ocean general circulation model (OGCM, grid point model as opposed to box models) rendering realistic results not only for large-scale budgets of heat and tracers but also for their three-dimensional time-dependent distributions? Yes, there was: Ernst Maier-Reimer (usually called "EMR", see Fig. 3.5) who took up an idea by Hasselmann (1982) of constructing a fast OGCM.

In order to achieve computational rather efficient OGCM suitable for climate studies, it should be sufficient to concentrate *"on spatial scales* (that) *are large compared with the internal Rossby radius throughout most of the ocean, while the characteristic time scales are large compared with the periods of gravity modes and barotropic Rossby wave modes. Under these scaling conditions, the baroclinic velocity field can be determined diagnostically from the given density field through*

[3] *"EMR" was the nickname of Ernst Maier-Reimer; "EMR" was quite commonly used at the MPI.*

[4] Christoph Heinze in part prepared a first version of this Section.

*local geostrophy. The barotropic velocity field and surface elevation can also be determined diagnostically for a given density field as the equilibrium barotropic response to the prescribed surface-stress forcing and bottom torque"* (Maier-Reimer et al. 1993).[5]

As estimated by Hasselmann (1982), a long time step of one month was possible for a horizontal resolution of 3.5° × 3.5° on an Arakawa E grid,[6] if the fast modes were filtered as mentioned above. I.e., a numerically fast ocean model was available for simulations covering millennia. However, the chosen filtering procedure resulted in several numerical instability problems (Maier-Reimer et al. 1993). Therefore, Maier-Reimer applied a more conventional method for filtering out the fast modes (Maier-Reimer et al. 1982): In particular, by omitting the non-linear advection of momentum in the primitive equations and by applying an implicit time integration scheme, a time step of one month could be obtained, while the model remained numerically stable and showed reliable results (Maier-Reimer et al. 1993).

The Large Scale Geostrophic (LSG) physical ocean general circulation model was established. Not only were the current systems and deep-water production adequately simulated, but also the water mass characteristics (temperature, salinity) and the sea ice distribution. The LSG model could be integrated into quasi-equilibrium from static initial conditions over a time frame of 2000 years in only 2–4 weeks of turnaround time on the then-existing supercomputer facilities (which were poorer than a standard laptop of today).

The general idea of the LSG modelling philosophy as a fast OGCM was again revived after the LSG was more or less abandoned at MPI in favour of higher resolution full "primitive" equation models. That idea, however, is followed-up in at least two Earth system Models of Intermediate Complexity (EMICs) (GENIE, Lenton et al. 2006; BERN3D, Ritz et al. 2011) though in a somewhat different modelling dialects, based on the ocean model components by Edwards and Shepherd (2002) and Müller et al. (2006). These EMICs are scientific backbone tools used in numerous applications, such as Monte-Carlo simulations for anthropogenic climate projections or in long-term paleoclimatic studies. For more realistic centennial-scale climate simulations with higher resolution, of course, more "nervous" primitive equation OGCMs[7] are needed. However, for fast multiple model integrations and also for obtaining system understanding by "playing" with multiple parameter perturbations, models with the LSG philosophy are still indispensable today and will be in future.

The LSG model was not the only model produced by Maier-Riemer. Rather, in his "desk" there was treasure of numerical models. For instance, Maier-Reimer provided the primitive equation "Äquatormodell" (equator model), potentially called "Flex", when Mojib Latif started his PhD work on simulating ENSO, resulting in papers such as Latif (1987) or Latif et al. (1988).

[5] The equatorial region required special treatment.

[6] Figure 5.9 provides an impression of the resolution.

[7] I.e., OGCMs that do not filter as strongly as LSG does.

## 5.3 Anomaly Coupling in Climate Modelling

### 5.3.1 *A Challenge: Coupling Atmosphere and Ocean General Circulation Models*

The current assessment reports of IPCC (e.g., IPCC 2021), in particular, their projections of the future development of climate for different scenarios of emissions of greenhouse gases and further climate forcers, are based on simulations with various comprehensive Earth system models. While such models are standard in today's climate modelling, the situation was quite different forty years ago in the 1980s.

At that time, both atmosphere and ocean general circulation models (GCMs) existed. The spatial resolution of these models was rather coarse (5–10° horizontally), and the number of implemented physical processes was rather limited. Nevertheless, it was obvious to couple such models in order to consistently simulate the joint atmosphere-ocean system within a single framework. This would allow to better understand the coupled system, including additional modes arising from the interaction of both spheres, the atmosphere and the ocean. This required technical solutions to various problems: E.g., the wind/currents velocities and the heat capacities in the atmosphere and the ocean are quite different, which resulted in rather different numerical time steps in atmospheric and oceanic GCMs and quite different time periods to reach a quasi-equilibrium (several orders of magnitude). Additionally, atmospheric and oceanic GCMs were constructed independently with rather different formulations of the boundary condition at the air-sea interface.

While the above-mentioned technical problems were overcome, a new, potentially unexpected challenge turned up: When the atmosphere and ocean GCMs, both in their individual quasi-equilibrium states, were coupled, the new atmosphere-ocean GCM gradually drifted to a new quasi-equilibrium state. As an example, Fig. 5.4 shows that the atmosphere temperature substantially differs in the coupled and uncoupled modes in the simulations by Washington et al. (1980). As can be seen, the zonal mean near-surface temperature for January is several °C lower in the coupled mode than in the atmosphere only mode. Analogous results were obtained by other modelling groups (e.g., Hansen et al. 1984; Gates et al. 1985).

### 5.3.2 *The Flux Correction Method*

The drift of coupled atmosphere-ocean GCMs can be overcome by the "flux correction method" (Sausen et al. 1988). The principal idea of this method was

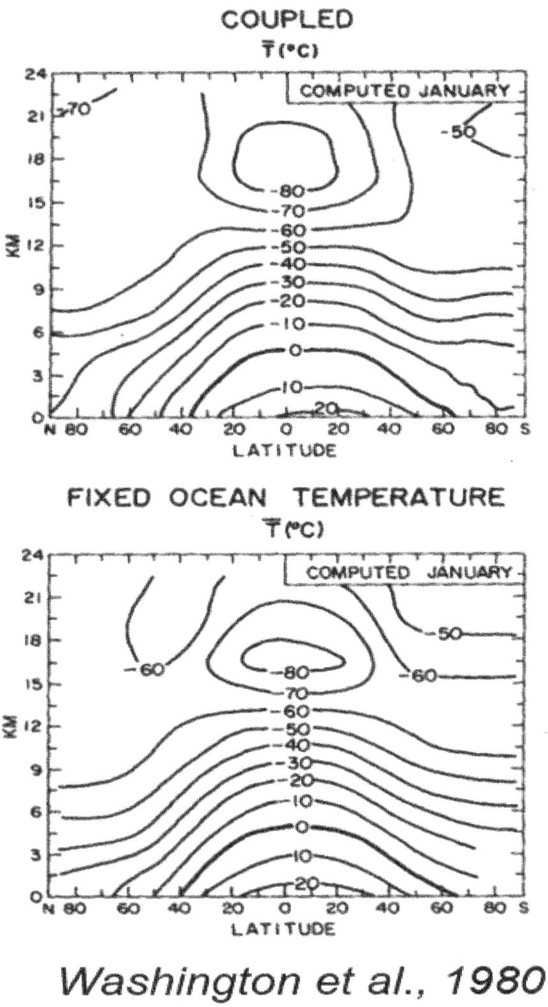


**Fig. 5.4** Zonal mean atmosphere temperature for January in the coupled mode (upper panel) and in the atmosphere only mode with prescribed sea surface temperature (lower panel). This Figure is based on a figure of Washington et al. (1980), © American Meteorological Society, used with permission

first presented at the Deutsche Meteorologen Tagung 1986 in Münster, Germany (Sausen 1986).[8]

"*The basic principle of the flux correction method is to couple the atmosphere and the ocean in such a manner that in the unperturbed case each subsystem simulates its own mean climate in the same manner as in the uncoupled mode, but responds fully interactively to the other sub-system in climate response or sensitivity experiments*" (Sausen et al. 1988). This is illustrated in Fig. 5.5. Here $\Phi$ and $\Psi$

[8] *Being a young post-doc at that time, Sausen was extremely nervous at the beginning of his presentation. Then he realized that the chairman, Prof. Hinzpeter, had forgotten to start the timer, and Sausen relaxed as the problem of staying within the assigned time was solved.*

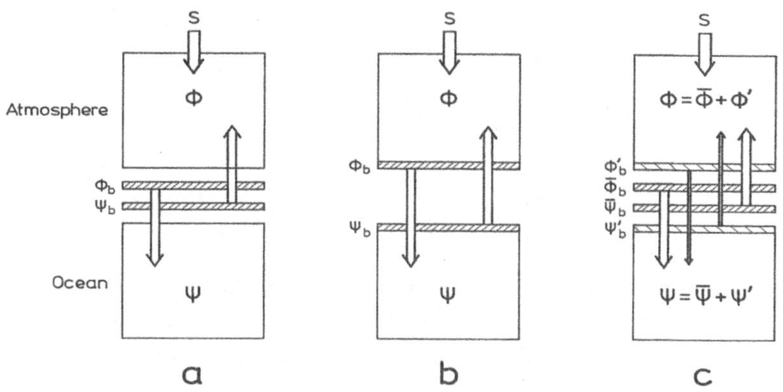


**Fig. 5.5** Boundary or coupling conditions of atmosphere and ocean models in different modes: (**a**) uncoupled; (**b**) fully coupled and (**c**) flux corrected (from Sausen et al. 1988)

denote the atmospheric and oceanic state vectors, resp., $\Phi_b$ and $\Psi_b$ are the boundary data required by the ocean and atmosphere model, respectively, in order to calculate the air-sea fluxes of momentum, heat and mass. In the uncoupled mode (panel a), the boundary data are taken offline from observed data. In the fully coupled mode (panel b) the boundary data are provided online from the other model. In the uncoupled mode, the prescribed boundary data ensure that both, the atmosphere and the ocean GCMs, stay close to the desired state of the model. In the (fully) coupled mode, the atmosphere-ocean model adjusts to its own quasi-equilibrium state. As a consequence, the air-sea fluxes of the uncoupled and coupled models are different.

In the flux corrected mode (Fig. 5.5, panel c) only the deviations ($\Phi_b'$ and $\Psi_b'$) from the mean states ($\overline{\Phi}_b$ and $\overline{\Psi}_b$) are online transferred to the respective other model component, while the mean states of the boundary data ($\overline{\Phi}_b$ and $\overline{\Psi}_b$) are from observations. The mean state may also include mean annual cycles. Consequently, applying this technique, the coupling of the air-sea fluxes is "corrected" (see below, and Sects. 5.3.3 and 5.3.4). By this method, we separate the drift problem from determining the response of the coupled system to changes in the climate forcing, e.g., a change in the $CO_2$ concentration or in the Earth's albedo.

In order to provide a more specific formulation of the flux correction method, we consider a consider a coupled atmosphere($\Phi$)-ocean($\Psi$) model described by the equations

$$\frac{\partial \Phi}{\partial t} = G_A(\Phi, t) + F(\Phi_b, \Psi_b, t) + E_A(\Phi, \Psi, t) \tag{5.1}$$

$$\frac{\partial \Psi}{\partial t} = G_O(\Psi, t) - F(\Phi_b, \Psi_b, t) + E_O(\Phi, \Psi, t) \tag{5.2}$$

where $G_A$ and $G_O$ are functions representing all physical processes occurring within the atmospheric and oceanic sub-systems, resp., $F$ is the flux from the ocean into the atmosphere, and $t$ is the time. The coupled model evolution differs from the true (or perfect) system by unknown model errors $E_A$ and $E_O$:

$$\frac{\partial \Phi}{\partial t} = \frac{\partial \Phi_{true}}{\partial t} + E_A(\Phi, \Psi, t) \tag{5.3}$$

$$\frac{\partial \Psi}{\partial t} = \frac{\partial \Psi_{true}}{\partial t} + E_O(\Phi, \Psi, t) \tag{5.4}$$

One of the central goals of atmosphere-ocean modelling has been to improve the models such that the errors $E_A$ and $E_O$ disappear or get so small that a remaining model drift becomes negligible. This challenge remains true for the latest generation of Earth system models.

There have been many attempts to heal the drift problem, e.g., by applying a Newtonian correction term, as used by Sarmiento and Bryan (1982). However, when doing this, the coupled model's response to perturbations unfortunately changes as was demonstrated by Sausen et al. (1988).

In the flux correction method, the errors $E_A$ and $E_O$ are compensated in the quasi-equilibrium state by constant additive flux terms $\Delta F_A$ and $\Delta F_O$:

$$\frac{\partial \Phi}{\partial t} = G_A(\Phi, t) + F(\Phi_b, \Psi_b, t) + E_A + \Delta F_A \tag{5.5}$$

$$\frac{\partial \Psi}{\partial t} = G_O(\Psi, t) - F(\Phi_b, \Psi_b, t) + E_O - \Delta F_O \tag{5.6}$$

The flux correction terms may also include a mean annual cycle. These terms are chosen such that the quasi-equilibrium of the coupled system is identical to the combined quasi-equilibrium of the uncoupled systems. This results in:[9]

$$\Delta F_A = F(\Phi_b^u, \Psi_b^m, t) - F(\Phi_b^u, \Psi_b^u, t) \tag{5.7}$$

$$\Delta F_O = F(\Phi_b^m, \Psi_b^u, t) - F(\Phi_b^u, \Psi_b^u, t) \tag{5.8}$$

Here $\Phi_b^u$ and $\Psi_b^u$ denote the (annual) mean boundary data of the uncoupled atmosphere and ocean models, resp., in the quasi-equilibrium states. Analogously $\Phi_b^m$ and $\Psi_b^m$ are the boundary data from atmospheric and oceanic, resp., measurements (observations).

Note that the flux correction terms $\Delta F_A$ and $\Delta F_A$ do not depend on the state of the coupled model: They only depend on the annual means or mean annual cycles of the observations and of the outputs of the uncoupled models. Consequently, the flux

[9] A more detailed recipe for calculating the flux correction terms is provided by Sausen and Cubasch (1989).

correction terms do probably not impact the coupled model's response to climate perturbation experiments (see, e.g., Sect. 5.3.3).

### 5.3.3 *First Applications*

In order to demonstrate that the above-described flux correction method obtains its objectives, Sausen et al. (1988) applied the techniques in a rather simple atmosphere-ocean model and in a more comprehensive one.

The simple model was constructed using linear terms, with one exception in the ocean part: The energy input into the ocean is a step function with a linear transition between constant levels.

$$C_A \frac{dT_A}{dt} = R_A - \lambda_A T_A + k(T_O - T_A) \tag{5.9}$$

$$C_O \frac{dT_O}{dt} = R_O - \lambda_O T_O - k(T_O - T_A) \tag{5.10}$$

Here $T_A$ and $T_O$ denote the atmospheric and oceanic temperatures, resp., $C_A$ and $C_O$ the heat capacities of each system, $R_A$ and $R_O$ the solar inputs (incl. the constant part of the radiative losses), $\lambda_A T_A$ and $\lambda_O T_O$ the radiative losses (linearized about reference temperatures), and $k(T_O - T_A)$ the energy flux from the ocean to the atmosphere. While $R_A$ is assumed to be constant, $R_O$ is defined as the nonlinear function

$$R_O(T_O) = \begin{cases} R_O^{(1)} & \text{for } T_O \leq T_O^{(1)} \\ R_O^{(1)} + \dfrac{T_O - T_O^{(1)}}{T_O^{(2)} - T_O^{(1)}}(R_O^{(2)} - R_O^{(1)}) & \text{for } T_O^{(1)} < T_O < T_O^{(2)} \\ R_O^{(2)} & \text{for } T_O^{(2)} \leq T_O \end{cases} \tag{5.11}$$

where $R_O^{(2)} > R_O^{(1)}$ and $T_O^{(2)} > T_O^{(1)}$. Equation (5.11) is a simple parameterization of the sea-ice - albedo feedback, where the ocean is completely ice-free for $T_O \geq T_O^{(2)}$ and completely ice-covered for $T_O \leq T_O^{(1)}$.

This simple model is ideal to illustrate and test the flux correction method.[10] Sausen et al. (1988) chose one set of parameters in Eqs. (5.9)–(5.11) for a "perfect" model (*PM*), representing the "observations" perfectly. Hence, we obtain the

[10] It turned out also to be a rather suitable tool for testing techniques for asynchronous and periodically synchronous coupling of models (Sausen and Voss 1996; Voss and Sausen 1996).

observed atmosphere and ocean temperatures $T_A^{(m)}$ and $T_O^{(m)}$ as the equilibrium state of the perfect model.

We assume that the parameters of the perfect model are unknown and that we have been able to "find" the parameters of an "approximate" model (*AM*), which also obeys Eqs. (5.9)–(5.11), but with parameters different from those in the perfect model (see Table 1 in Sausen et al. 1988). Actually, the parameters of *PM* and *AM* differ only in $\lambda_A$ and $\lambda_O$ by about 1%.

In the uncoupled mode, both the atmosphere approximate model and the ocean approximate model, resp., reproduce the temperature observations, i.e., the coupled perfect model temperatures, with a bias of up to 0.2 K (see also Table 2 of Sausen et al. 1988). Nevertheless, the ocean-atmosphere fluxes of the uncoupled approximate models differ by 3.23 W/m$^2$, about 4%, from *PM*.

Due to this difference in the coupling fluxes, the coupled approximated model *AM* drifts from the combined equilibrium temperatures of the uncoupled models to a new state, which is nearly 10 K warmer than the equilibrium state of the perfect model *PM*. The solid curve in Fig. 5.6 illustrates this drift. In terms of the simple models, this means that *AM* drifts from an ice-covered to an ice-free state. This would have consequences for a potential climate warming experiment: As the sea ice is already gone due to model drift, a simulated climate warming would be smaller (see below). This could be interpreted as a smaller climate sensitivity.

Following Eqs. (5.7) and (5.8), the flux correction terms of our simple model are:

$$\Delta F_A = +k(T_O^{(m)} - T_O^{(u)1}) \tag{5.12}$$

$$\Delta F_O = -k(T_A^{(m)} - T_A^{(u)1}) \tag{5.13}$$

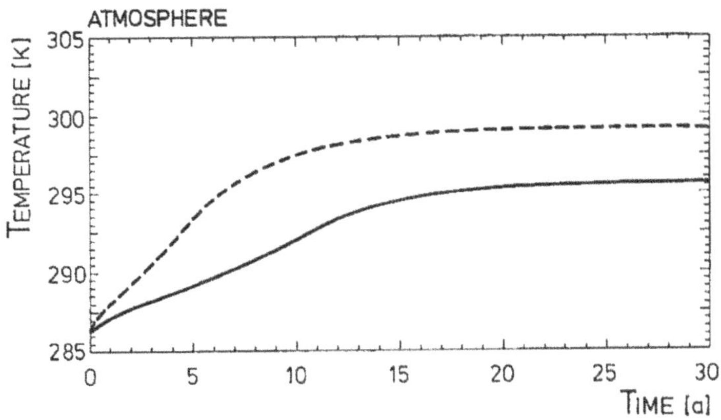


**Fig. 5.6** Time evolution of the atmosphere temperature of the coupled approximate model *AM* without flux correction for the control run (solid) and the reduced emissivity run (dashed) (from Sausen et al. 1988)

Where $T_A^{(u)}$ and $T_O^{(u)}$ denote the equilibrium atmosphere and ocean temperatures of the uncoupled approximate models, respectively. With these flux correction terms *AM* takes the form

$$C_A \frac{dT_A}{dt} = R_A - \lambda_A T_A + k(T_O - T_A) + k(T_O^{(m)} - T_O^{(u)]}) \tag{5.14}$$

$$C_O \frac{dT_O}{dt} = R_O - \lambda_O T_O - k(T_O - T_A) - k(T_A^{(m)} - T_A^{(u)]}) \tag{5.15}$$

With flux correction, the coupled approximate model *AM* no longer drifts into a new state when starting from the equilibrium solutions of the uncoupled models. In Fig. 5.7, this is illustrated by the solid curve for the atmosphere temperature.

Having demonstrated that the flux correction prevents the drift in the simple model *AM*, the question arises: what is the impact of the flux correction on the climate response to a (external) perturbation? To this end an experiment was performed where the emissivities of the atmosphere $\lambda_A$ in Eqs. (5.9) and (5.14) was reduced by 2% each. This might be interpreted as the effect of additional greenhouse gases. The impact on the atmosphere temperature of the coupled model *AM* without and with flux correction is illustrated by the dashed curves in Figs. 5.6 and 5.7, respectively. Without flux correction, the warming is ≈3.5 K, while the warming is ≈9.2 K with flux correction (see also Table 2 of Sausen et al. 1988). Contrary to more complex models, we know the "true" response in the case of our simple model: ≈9.2 K, as a result of the perfect model *PM*.

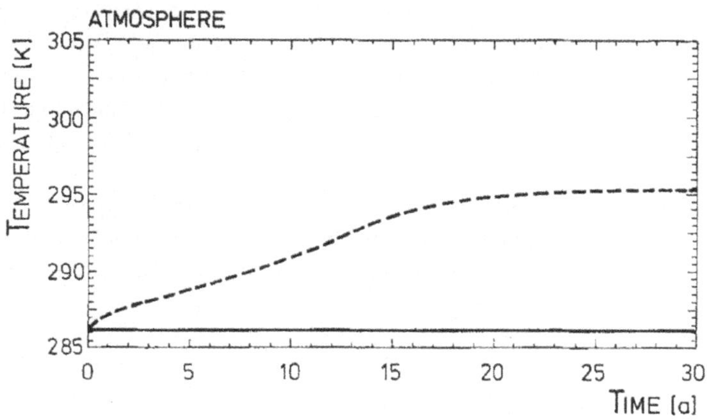


**Fig. 5.7** Time evolution of the atmosphere temperature of the coupled approximate model *AM* with flux correction for the control run (solid) and the reduced emissivity run (dashed) (from Sausen et al. 1988)

In a second example, Sausen et al. (1988) applied the flux correction method to a more comprehensive atmosphere-ocean model.[11] The ocean model was the LSG (large-scale geostrophic) model (Sect. 5.2) by Maier-Reimer et al. (1983), see also Maier-Reimer (1985) and Maier-Reimer and Hasselmann (1987). The model had a horizontal resolution of $3.54° \times 3.54°$ on an Arakawa E-grid and five vertical levels. A numerical filtering of gravity waves allowed an implicit time step of one month.

The atmosphere model was a two-layer radiative-convective model (Sausen 1989) with the same horizontal resolution as the ocean model The horizontal transport of heat is performed through temperature advection by the observed wind and a (weak) diffusive transport. The time step was 1 hour.[12]

The models were coupled by the total air-sea heat flux. Due to the rather large ocean time, the atmosphere model was in a quasi-diagnostic mode, i.e., after 240 atmosphere time steps (=10 days), a new equilibrium was achieved. The long time step, rather coarse spatial resolution, the small number of prognostic variables, and the limited physical processes allowed us to perform several centennial simulations at the computing capacity on a CYBER205 available in Hamburg during the late 1980s.

Due to large differences in the air-sea heat flux used by the uncoupled atmosphere and ocean models, resp., the coupled model without flux correction exhibited a severe drift. Within 100 years the global mean atmosphere temperature at all levels was reduced by $\approx 3$ K, while the global mean temperature at the uppermost ocean level decreased by $\approx 5$ K (see dashed lines in Fig. 5.8 for the atmosphere temperature). With flux correction, the drift was avoided for both the atmosphere (Fig. 5.8 full lines) and the ocean.

For illustration Fig. 5.9 shows the temperature drift in the ocean over 100 years at 75 m depth (uppermost ocean layer). Without flux correction, the range of drifts exceeds the interval from $-16$ K to +6 K, and the spatial structure is rather inhomogeneous (panel a). With flux correction, a small remaining drift is only found in the Norwegian Sea (panel b). This remaining drift might be attributed to a drift of the uncoupled ocean model (panel c), which potentially had not yet achieved its quasi-equilibrium when the coupled run was started.

Sausen et al. (1988) also performed a "climate response" experiment with the more comprehensive coupled model: the absorption coefficients for the long wave radiation were slightly increased. The responses without and with flux correction were quite different.

[11] Presumably, this was the first coupled "comprehensive" atmosphere-ocean model run in Hamburg.

[12] Klaus Hasselmann named this model "a poor man's atmosphere" as it only required rather small computational power.

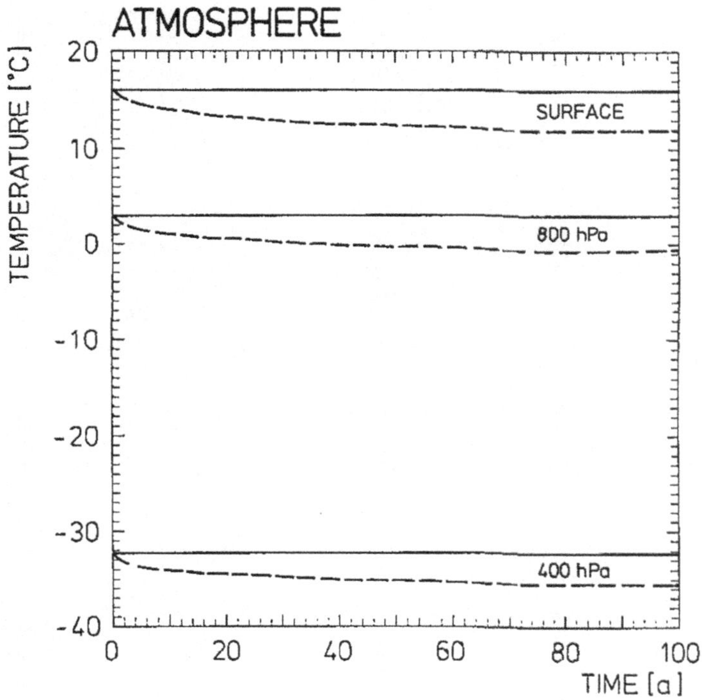


**Fig. 5.8** Time evolution of the global mean atmosphere temperature of the coupled "comprehensive" model without (dashed) and with (solid) flux correction for a control run (from Sausen et al. 1988)

## *5.3.4 Flux Correction Versus Anomaly Coupling*

Naming the above-described modelling technique "flux correction" caused quite a controversial discussion. For some people, the term "flux correction" induced the feeling that something tricky has been done to coupled climate models.[13] Potentially less excitement would have been triggered if Sausen et al. (1988) had chosen the

[13] *Sausen presented the flux correction to an international audience during the short evening presentations of the students at NATO Advanced Study Institute (a summer school) on "Physically Based Modelling and Simulation of Climate and Climatic Change" in Erice/Italy, 12–23 May 1986. While Sausen's presentation was limited to 15 min, the discussion went on for about 1 hour. It was not so much a discussion with the author of the presentation (a post-doc at that time), rather two groups of the lectures, international climate modellers with a high reputation, discussed the method quite controversially and intensively. On the one hand side, the Americans were of the opinion that such a method was not physically justified and, hence, should not be used. On the other hand, the Europeans strongly supported the technique, which could help overcome a severe problem in coupled atmosphere-ocean modelling. The young post-doc Sausen was between the "fronts". But he felt quite protected by Klaus Hasselmann in his role as a godfather.*

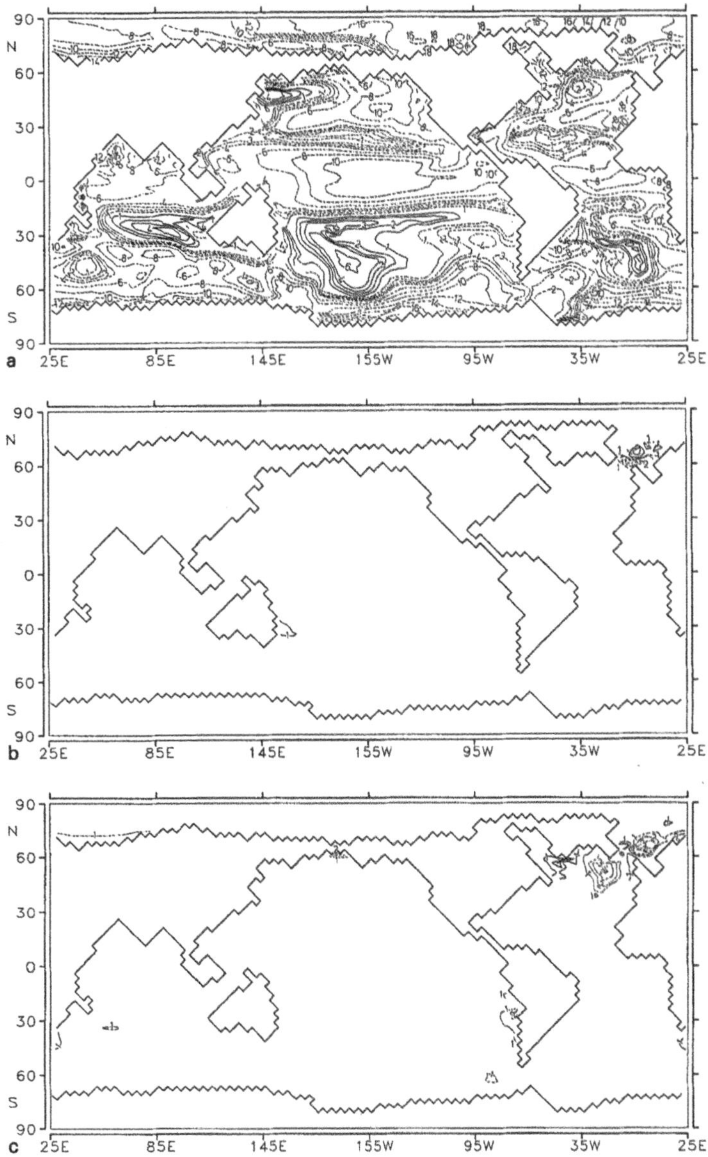


**Fig. 5.9** Temperature drift in the ocean over 100 years at 75 m depth: (**a**) without flux correction, (**b**) with flux correction, and (**c**) for the uncoupled ocean. Negative isolines are dashed; the zero line is not plotted (from Sausen et al. 1988)

term "anomaly coupling" as suggested by Fig. 5.5 and illustrated by Section 2 of Sausen et al. (1988).

This can made more obvious by a reformulation of Eqs. (5.5)–(5.8) (see also Sausen and Cubasch 1989): The boundary values $\Phi_b$ and $\Psi_b$ are split in the contributions from the uncoupled model and the deviations from the uncoupled contributions:

$$\Phi_b = \Phi_b^u + \delta\Phi_b \tag{5.16}$$

$$\Psi_b = \Psi_b^u + \delta\Psi_b \tag{5.17}$$

Putting this into Eq. (5.5) with Eq. (5.7) reveals:

$$\frac{\partial\Phi}{\partial t} = G_A(\Phi, t) + F(\Phi_b^u + \delta\Phi_b, \Psi_b^u + \delta\Psi_b) + F(\Phi_b^u, \Psi_b^m) - F(\Phi_b^u, \Psi_b^u) \tag{5.18}$$

Here, we omit the error term and only display the equations for the atmosphere. The ocean part can be treated analogously.

We now expand the blue term in Eq. (5.18) into a Taylor series:

$$\begin{aligned}\frac{\partial\Phi}{\partial t} = {} & G_A(\Phi, t) \\ & + F(\Phi_b^u, \Psi_b^u) \\ & + \delta\Phi_b \frac{\partial}{\partial\, \delta\Phi_b} F(\Phi_b^u + \delta\Phi_b, \Psi_b^u + \delta\Psi_b) \\ & + \delta\Psi_b \frac{\partial}{\partial\, \delta\Psi_b} F(\Phi_b^u + \delta\Phi_b, \Psi_b^u + \delta\Psi_b) \\ & + \text{terms of higher order} \\ & + F(\Phi_b^u, \Psi_b^m) - F(\Phi_b^u, \Psi_b^u)\end{aligned} \tag{5.19}$$

Finally, we obtain

$$\begin{aligned}\frac{\partial\Phi}{\partial t} \approx {} & G_A(\Phi, t) + F(\Phi_b^u, \Psi_b^m) \\ & + \delta\Phi_b \frac{\partial}{\partial\, \delta\Phi_b} F(\Phi_b^u + \delta\Phi_b, \Psi_b^u + \delta\Psi_b) \\ & + \delta\Psi_b \frac{\partial}{\partial\, \delta\Psi_b} F(\Phi_b^u + \delta\Phi_b, \Psi_b^u + \delta\Psi_b)\end{aligned} \tag{5.20}$$

Hence, the flux correction turns into a coupling that is linear in anomalies of the air-sea flux, i.e., we have an anomaly coupling.[14]

[14] There are quite a few papers where the name "flux adjustment" is used for the flux correction.

### 5.3.5 Final Remarks

The last paragraph of Sausen et al. (1988) reads: "*Finally, we note that the flux correction method is useful not only as a technique for maintaining the coupled model at the "working point" for which the model sub-systems were designed but also as a diagnostic tool. The flux correction fields show the structure of the mutual inconsistencies of the separate sub-systems, indicating where modifications in the separate sub-systems are most needed.*

Shortly after the original publication of the flux correction method (Sausen et al. 1988), several extensions and further illustrations of the method were published, e.g., Lunkeit and Sausen (1989), Sausen and Lunkeit (1989a,b, 1990) (Fig. 5.10).

The flux correction method was a necessary tool that enabled the first Hamburg comprehensive coupled atmosphere-ocean models, where both components were general circulation models, both state of the art in the 1990s (Cubasch et al. 1992; Lunkeit et al. 1996). Also later, the flux correction method has found reception in the peer-reviewed literature (Fig. 5.11) with a peak in 1996, and even after 25 years, the method has been referenced.

Many of the coupled models of intermediate complexity that provided information for the Fifth Assessment Report of IPCC (Flato et al. 2013) used flux correction (here called "flux adjustment"): 15 out of 15 models (Table 9.A.2 of Flato et al.

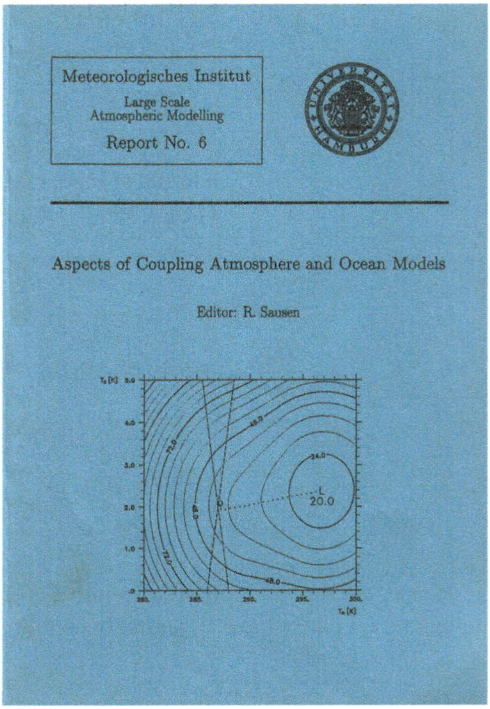


**Fig. 5.10** Internal Report of the Meteorologisches Institut, Universität Hamburg, Germany, published in 1989

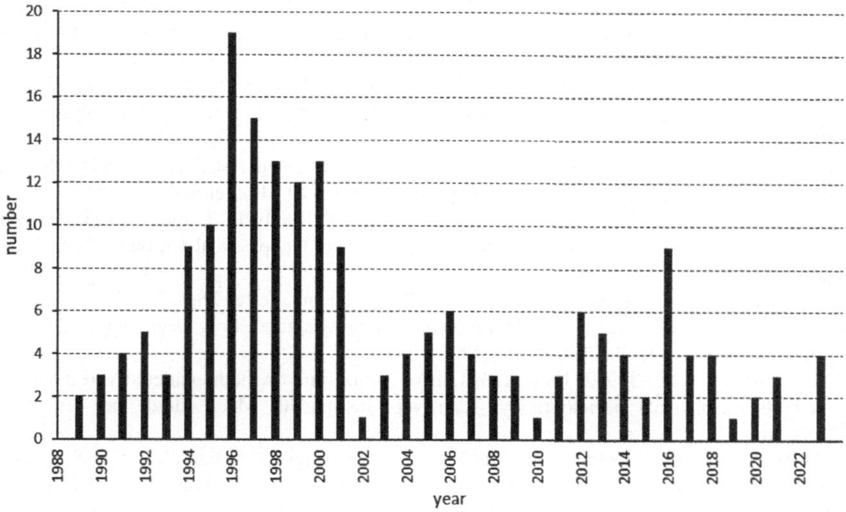


**Fig. 5.11** Annual citations of Sausen et al. (1988) based on scopus.com (downloaded at 17 Dec 2023)

2013). Unfortunately, an analogous table (Table 9.A.1 of Flato et al. 2013) on the more comprehensive models does not provide information on the coupling and whether a flux correction was used. Similarly, this information is not provided by the Sixth Assessment Report of IPCC (IPCC 2021).

In the recent year, Sausen et al. (1988) was cited by four papers on rather different modelling aspects: Lian (2023), Masunaga et al. (2023), Schevenhoven et al. (2023), and Shutyaev et al. (2023).

# References

Cubasch U, Hasselmann K, Höck H, Maier-Reimer E, Mikolajewicz U, Santer BD, Sausen R (1992) Time-dependent greenhouse warming computations with a coupled ocean-atmosphere model. Clim Dyn 8:55–69

Edwards N, Shepherd J (2002) Bifurcations of the thermohaline circulation in a simplified three-dimensional model of the world ocean and the effects of inter-basin connectivity. Clim Dyn 19(1):31–42. https://doi.org/10.1007/s00382-001-0207-7

Fischer G (1959) Ein numerisches Verfahren zur Errechnung von Windstau und Gezeiten in Randmeeren. Tellus 11(1):60–76. https://doi.org/10.3402/tellusa.v11i1.9281

Flato G, Marotzke J, Abiodun B, Braconnot P, Chou S, Collins W, Cox P, Driouech F, Emori S, Eyring V, Forest C, Gleckler P, Guilyardi E, Jakob C, Kattsov V, Reason C, Rummukainen M (2013) Evaluation of climate models. In: Stocker T, Qin D, Plattner GK, Tignor M, Allen S, Boschung J, Nauels A, Xia Y, Bex V, Midgley P (eds) Climate change 2013: the physical science basis. Contribution of working group I to the fifth assessment report of the intergovernmental panel on climate change. Cambridge University Press, Cambridge, pp 741–866. https://doi.org/10.1017/9781009157896

Gates WL, Han YJ, Schlesinger ME (1985) The global climate simulated by a coupled atmosphere-ocean general circulation model: preliminary results. Elsevier Oceanogr Ser 40(C):131–151. https://doi.org/10.1016/S0422-9894(08)70707-4

Hansen W (1956a) A simple model of oceanic circulation; [Ein einfaches Modell der Zirkulation im Meere]. Deutsche Hydrographische Zeitschrift 9(2):102–106. https://doi.org/10.1007/BF02021148

Hansen W (1956b) Theorie zur Errechnung des Wasserstandes und der Strömungen in Randmeeren nebst Anwendungen. Tellus 8(3):287–300. https://doi.org/10.3402/tellusa.v8i3.9023

Hansen J, Lacis A, Rind D, Russell G, Stone P, Fung I, Ruedy R, Lerner J (1984) Climate sensitivity: analysis of feedback mechanisms. American Geophysical Union (AGU). https://doi.org/10.1029/gm029p0130

Hasselmann K (1976) Stochastic climate models. Part I. Theory. Tellus 28:473–485

Hasselmann K (1982) An ocean model for climate variability studies. Prog Oceanogr 11(2):69–92. https://doi.org/10.1016/0079-6611(82)90004-0

Institut für Meereskunde (1962) Proceedings of the symposium on mathematical-hydrodynamical methods of physical oceanography. Tech. rep. Institut für Meereskunde der Universität Hamburg

IPCC (2021) In: Masson-Delmotte V, Zhai P, Pirani A, Connors SL, Péan C, Berger S, Caud N, Chen Y, Goldfarb L, Gomis MT, Huang M, Leitzell K, Lonnoy E, Matthews JBR, Maycock TK, Waterfield T, Yelekçi O, Yu R, Zhou B (eds) Climate change 2021: the physical science basis. Contribution of working group I to the sixth assessment report of the intergovernmental panel on climate change. Cambridge University Press, Cambridge, 2391 pp. https://doi.org/10.1017/9781009157896

Latif M (1987) Tropical ocean circulation experiments. J Phys Oceanogr 17:246–263

Latif M, Biercamp J, Von Storch H (1988) The response of a coupled ocean-atmosphere general circulation model to wind bursts. J Atmos Sci 45:964–979

Lenton TM, Williamson MS, Edwards NR, Marsh R, Price AR, Ridgwell AJ, Shepherd JG, Cox SJ, team TG (2006) Millennial timescale carbon cycle and climate change in an efficient earth system model. Clim Dyn 26(7):687–711. https://doi.org/10.1007/s00382-006-0109-9

Lian T (2023) ENSO spread caused by uncertainties in initial condition and high-frequency perturbations in the Tropical Pacific. J Clim 36(15):5143–5161. https://doi.org/10.1175/JCLI-D-22-0658.1

Lunkeit F, Sausen R (1989) Flux correction with annual cycle and stochastic forcing. In: Sausen R (ed) Aspects of coupling atmosphere and ocean models. Large Scale Atmospheric Modelling Report, vol 6. Meteorologisches Institut der Universität Hamburg, Hamburg, pp 1–15

Lunkeit F, Sausen R, Oberhuber JM (1996) Climate simulations with the global coupled atmosphere-ocean model ECHAM2/OPYC. Part I: present-day climate and enso events. Clim Dyn 12(3):195–212. https://doi.org/10.1007/BF00211618

Maier-Reimer E (1985) A large scale ocean circulation model. Techn. Report. Max-Planck-Institut für Meteorologie, Hamburg

Maier-Reimer E, Hasselmann K (1987) Transport and storage of $CO_2$ in the ocean - an inorganic ocean-circulation carbon cycle model. Clim Dyn 2(2):63–90

Maier-Reimer E, Müller D, Olber D, Willebrand J, Hasselmann K (1982) Ein Modell der ozeanischen Zirlulation zur Untersuchung von Klimaschwankungen. Max-Planck-Institut für Meteorologie, Hamburg. https://hdl.handle.net/21.11116/0000-0003-A552-6

Maier-Reimer E, Hasselmann K, Olbers D, Willebrand J (1983) An ocean circulation model for climate studies. Techn. Report. Max-Planck-Institut für Meteorologie, Hamburg

Maier-Reimer E, Mikolajewicz U, Hasselmann K (1993) Mean circulation of the hamburg LSG OGCM and its sensitivity to the thermohaline surface forcing. J Phys Oceanogr 23(4):731–757

Masunaga R, Miyakawa T, Kawasaki T, Yashiro H (2023) Flux adjustment on seasonal-scale sea surface temperature drift in NICOCO. J Meteorol Soc Jpn 101(3):175–189. https://doi.org/10.2151/jmsj.2023-010

Müller P, von Storch H (2004) Computer modelling in atmospheric and oceanic sciences - building knowledge. Springer, Berlin

Müller SA, Joos F, Edwards NR, Stocker TF (2006) Water mass distribution and ventilation time scales in a cost-efficient, three-dimensional ocean model. J Clim 19(21):5479–5499. https://doi.org/10.1175/JCLI3911.1

Ritz SP, Stocker TF, Joos F (2011) A coupled dynamical ocean–energy balance atmosphere model for paleoclimate studies. J Clim 24(2):349–375. https://doi.org/10.1175/2010JCLI3351.1

Sarmiento J, Bryan K (1982) Ocean transport model for the North Atlantic. J Geophys Res 87(C1):394–408. https://doi.org/10.1029/JC087iC01p00394

Sausen R (1986) Gekoppelte Ozean-Atmosphären-Modelle ohne Drift der prognostischen Variablen. Ann Meteorol (neue Folge) 23:82–83

Sausen R (1989) A "poor man's atmosphere" – a radiative-convective model with advection by prescribed winds. Large Scale Atmospheric Modelling Report, vol 2. Meteorologisches Institut der Universität Hamburg, Hamburg

Sausen R, Cubasch U (1989) A technical guide for the flux corrections method. In: Sausen R (ed) Aspects of coupling atmosphere and ocean models. Large scale atmospheric modelling report, vol 6. Meteorologisches Institut der Universität Hamburg, Hamburg, pp 39–71

Sausen R, Lunkeit F (1989a) A modified flux correction method for unstationary models. In: Sausen R (ed) Aspects of coupling atmosphere and ocean models. Large scale atmospheric modelling report, vol 6. Meteorologisches Institut der Universität Hamburg, Hamburg, pp 17–28

Sausen R, Lunkeit F (1989b) Illustrating the climate drift of coupled ocean-atmosphere models by means of a potential. In: Sausen R (ed) Aspects of coupling atmosphere and ocean models. Large scale atmospheric modelling report, vol 6. Meteorologisches Institut der Universität Hamburg, Hamburg, pp 29–37

Sausen R, Lunkeit F (1990) Some remarks on the cause of the climate drift of coupled ocean-atmosphere models. Beitr Phys Atmos 63:141–146

Sausen R, Voss R (1996) Techniques for asynchronous and periodically synchronous coupling of atmosphere and ocean models: Part I: general strategy and application to the cyclo-stationary case. Clim Dyn 12:313–323. https://doi.org/10.1007/BF00231105

Sausen R, Barthel K, Hasselmann K (1988) Coupled ocean-atmosphere models with flux correction. Clim Dyn 2(3):145–163. https://doi.org/10.1007/BF01053472

Schevenhoven F, Keenlyside N, Counillon F, Carrassi A, Chapman WE, Devilliers M, Gupta A, Koseki S, Selten F, Shen ML, Wang S, Weiss JB, Wiegerinck W, Duane GS (2023) Supermodeling improving predictions with an ensemble of interacting models. Bull Am Meteorol Soc 104(9):E1670–E1686. https://doi.org/10.1175/BAMS-D-22-0070.1

Shutyaev V, Zalesny V, Agoshkov V, Parmuzin E, Zakharova N (2023) Four-dimensional variational data assimilation and sensitivity of ocean model state variables to observation errors. J Mar Sci Eng 11(6). https://doi.org/10.3390/jmse11061253

Sündermann J (1989) Walter Hansen - zum 80ten Geburtstag. DGM Mitteilungen 3:16–17

Sündermann J, Vollmers H (1972) Tidewellen in Ästuarien. Wasserwirtschaft 62:1–9

Voss R, Sausen R (1996) Techniques for asynchronous and periodically synchronous coupling of atmosphere and ocean models: Part II: impact of variability. Clim Dyn 12:605–614. https://doi.org/10.1007/bf00216269

Washington W, Semtner A, Meehl G, Knight D, Mayer T (1980) A general circulation experiment with a coupled atmosphere, ocean and sea ice model. J Phys Oceanogr 10(12):1887–1908. https://doi.org/10.1175/1520-0485(1980)010<1887:agcewa>2.0.co;2

# Chapter 6
# Downscaling

**Hans von Storch, Eduardo Zorita, Robert Sausen, Martin Claussen, and Martin Heimann**

**Abstract** The primary goal of dynamical climate models is to create future projections, reconstruct past conditions, and perform numerical experiments. These models, whether global or regional, are most accurate at large scales, but their accuracy diminishes at smaller scales. This decline in skill is partly because noise becomes increasingly significant at smaller scales, and also because the models must inevitably truncate the dynamics, ignoring states that aren't captured by the numerical discretization. However, many studies focusing on the impacts of climate variations and changes require an understanding of these smaller scales. As a result, there's a trend toward refining the models to capture smaller and smaller scales, though this approach significantly increases computational costs. A more economical alternative is to identify the statistical or dynamical relationships between large- and small-scale states, a process known as "downscaling." In this chapter, we explore the development of both empirical and dynamical downscaling methods for estimating consistent small-scale dynamics and assessing regional and local impacts. One of the technical challenges in this process is effectively distinguishing between small and large scales through appropriate filtering techniques.

The downscaling concept, both with its empirical and its dynamic variants, describes regional statistics as conditioned by large scales and allows for the reconstruction of past regional variability and regionalised global climate change projections.

In Hamburg, the issue got some attention when efforts to build and couple global models of the atmosphere and ocean matured, and were used to estimate climate change on the global and the continental scales under possible future atmospheric greenhouse-gas concentrations (Cubasch et al. 1992). Unfortunately, these global models could hardly describe significant climate impact variables, mostly of smaller spatial scale—particularly spatial distributions of precipitation (see Sect. 6.1). Then Klaus Hasselmann and his lieutenants began to ask which impacts would be associated with changes as described in such scenarios. In this chapter, we mostly describe concepts and results obtained by Hamburg climate scientists when trying to answer those questions.



H. von Storch et al., *Climate Science Concepts Born in Hamburg*,
https://doi.org/10.1007/978-3-031-81108-1_6

In principle, there are two approaches for estimating local and regional detail from the large-scale state. One is encapsulated in Victor Starr's (1942) understanding of the weather forecast problem: "*The general problem of forecasting weather conditions may be subdivided conveniently into two parts. In the first place, it is necessary to predict the state of motion of the atmosphere in the future; secondly, it is necessary to interpret this expected state of motion in terms of the actual weather it will produce at various localities. The first of these problems is essentially of a dynamic nature, inasmuch it concerns itself with the mechanics of the motion of a fluid. The second problem involves a large number of details because, under exactly similar conditions of motion, different weather types may occur, depending upon the temperature of the air involved, the moisture content of the air, and a host of local influences.*"

Following this concept, methods were built that determined the dynamic link between large-scale states and small-scale regional or local dynamic states. Obviously, this concept could only consider those regional and local variables that are directly linked to the conditioning large-scale variables, as in the case of large-scale wind patterns and regional and local precipitation. The method can hardly be applied to two variables that are statistically, but not dynamically coupled, such as large-scale seasonal mean circulation and the intraseasonal statistics of storm surge heights (see Sect. 6.3.2), or seasonal mean temperature and the timing of plant flowering (Maak and von Storch 1997).

As outlined in Starr's framework, it is unrealistic to anticipate a singular regional or local state that exclusively aligns with the state of the steering variable(s). Therefore, the process of downscaling essentially yields a collection of consistent regional and local states, with most methodologies opting for the expected average value within this ensemble.

This led to the concept of weather typing, i.e., of determining typical weather situations, or Grosswetterlagen, which was done with the statistical-dynamical downscaling (e.g., Frey-Buness et al. 1995; Fuentes and Heimann 2000).[1] From the change in the frequency of Grosswetterlagen, one would derive a change in the statistics of impacts, which typically go with the Grosswetterlagen.

The other, earlier, approach was first tried by Kim et al. (1984), namely, to relate statistics of large-scale states to statistics of small-scale states. In the spirit of Hasselmann's separation into a dynamically active signal (large-scale) part and a responding, mostly passive noise (small-scale) part (see Chap. 2), von Storch et al. (1993)[2] implemented this approach, who coined the term "downscaling" for relating large time-mean fields of one sort (here: monthly mean North Atlantic air pressure) to a regional time-mean field (here: monthly mean Iberian Peninsula

---

[1] Fuentes and Heimann summarize: "*The method is based on the disaggregation of a multi-year time-series of large-scale meteorological data into multi-day episodes of quasi-stationary circulation. The episodes are subsequently grouped into a defined number of classes. A regional model is used to simulate the evolution of weather during the most typical episode of each class.*"

[2] The concept and the example were first published as MPI Report 64, in 1991, see Fig. 6.1.

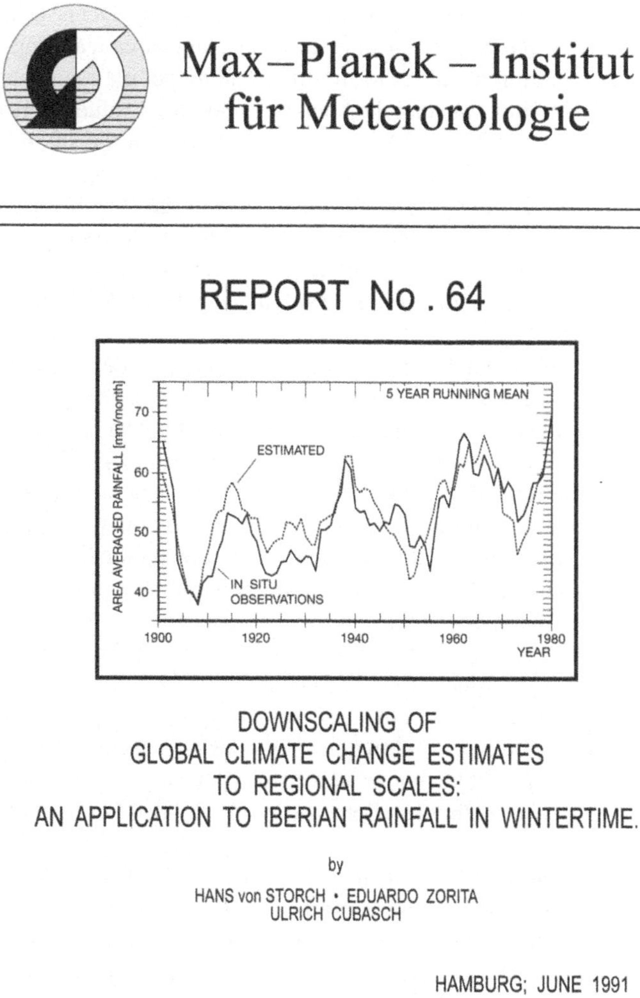
Max–Planck – Institut
für Meterorologie

REPORT No . 64



DOWNSCALING OF
GLOBAL CLIMATE CHANGE ESTIMATES
TO REGIONAL SCALES:
AN APPLICATION TO IBERIAN RAINFALL IN WINTERTIME.

by

HANS von STORCH • EDUARDO ZORITA
ULRICH CUBASCH

HAMBURG; JUNE 1991

**Fig. 6.1** Creating a terminology in 1991—"downscaling"

precipitation), thus a link between statistics of state variables, not between state variables themselves. Obviously, such downscaling links do not represent direct causal dynamic links, but a co-variation, which may need a chain of several causally motivated links.

Parallel to these empirical approaches, there were efforts to formulate regional atmospheric models run with boundary value-forcing (e.g., Giorgi and Mearns 1991). After overcoming the problem that the features inside the regional area on a finer grid would move inconsistently with the propagation of the same weather stream on the coarse grid, which determines the boundary values (Davies 1976), these models were used extensively,

An important step was the "big-brother experiment" by Denis et al. (2002b): Here, the output of a high-resolution simulation on a large area featuring a front was used as boundary and initial values to run the same model on a smaller area, with boundary values. The trick was that the boundary and initial values were coarsened, and the question was whether the high-resolution model on the small area ("little brother") would reproduce the development of the model on the larger area ("big brother") , even though only a coarse version of state of the "big brother" was given to the "little brother". After a short initial phase, the "little brother" was indeed capable of doing so, which demonstrated the skill of embedding a limited area model in a coarse environment in adding realistic small-scale detail if the region is not too large (cf. Fig. 6.17).

However, this conventional setup with a limited area model does not represent real downscaling, as it does not lead to processing large-scales to obtain smaller scales. This processing of large-scale fields can be done by what nowadays is mostly named "spectral nudging". After a first brief introduction of this concept by Waldron et al. (1996), von Storch et al. (2000) demonstrated the success of this approach in postprocessing coarse-grid multi-decadal re-analyses and climate change scenarios. This large-scale constraining guides the model in following the large-scale steering of the reanalyses better and reducing the internal variability in the interior of the model domain. (Details in Sect. 6.3.)

Earlier efforts to implement a top-down scale steering in limited area models of atmospheric models were tried out by Kida et al. (1991).

In the meantime, the concept, both empirical and dynamic downscaling, has been placed solidly in the toolbox of climate science, as is demonstrated by a variety of review papers, such as Takayabu et al. (2016), Benestad (2016), or Wang et al. (2004), and at least two textbooks (Benestad et al., 2008; Maraun and Widmann, 2018).

In the following Sect. 6.1 we sketch ideas on how to determine and separate spatial scales in geophysical fields. In Sect. 6.2, the success of global models on the large scales and the failure on small scales are illustrated, using an example from the 1990s. Obviously, the spatial grid resolution has been refined quite a bit in recent years, so the terms "small scales" and "large scales" have a different meaning nowadays and will continue to do so in future times.

In Sect. 6.3, the basic idea of *empirical downscaling*, as developed in Hamburg, is described and illustrated by several examples using linear regression methods (von Storch et al. 1993). Alternatives to regression, also suggested and developed in Hamburg, are analogues (Zorita and von Storch 1999) and kriging (Biau et al. 1999).

The *dynamic downscaling*-Sect. 6.4 first deals with the conventional approach of limited area modeling embedded in (most often) global analysis or simulation data (Sect. 6.4.1). The method developed and made popular in Hamburg is based on "spectral nudging". This section is divided into a Sect. 6.4.2 on limited area models and another brief Sect. 6.4.3 on global models. Where possible, cases involving oceanic variables and atmospheric variables are considered.

## 6.1 Scales: Determination and Separation

The simplest way to determine the scales of a geophysical field is to inspect it visually. However, in most cases, such fields are made up of contributions of a variety of scales, large-, medium-, and small scales. Then several methods are available to determine dominant scales, or to separate large- and small-scales:

1. A straight-forward way is *smoothing*: If, for simplicity, $f$ is a one-dimensional field on a regular grid, then a running mean may be determined:

$$M(f)_i = \frac{1}{2N+1} \sum_{k=-N,\dots,N} f_{i+k} \tag{6.1}$$

If $N$ is sufficiently large, then $M(f)$ will consists of the large-scale component of $f$, and $f - M(f)$ will show the small scale components.

2. A more sophisticated way is to expand $f$ into a series of orthonormal functions $g_i$

$$M(f) = \sum_i \alpha_i(f) g_i \tag{6.2}$$

with coefficients $\alpha_i(f) =< g_i, f >$, with $<, >$ representing the scalar product. If we again assume regular grids and $f$ is periodic or almost periodic, then the best-known case is trigonometric functions, i.e., sine and cosine functions—then it is called a *Fourier-expansion*; when $f$ is not periodic, then a cosine expansion may be appropriate (Denis et al. 2002a). For two dimensions of the sphere, spherical harmonics may be used. Since these functions go with a "wavelength", they are ordered according to scale.

Also, Empirical Orthogonal Functions (EOFs) may serve this purpose if their "natural ordering," following their relative importance in presenting variance of the ensemble of $f$-fields, is also an ordering of spatial scales. This ansatz is useful if the field $f$ is not on a regular lat-long grid. This application was suggested and tested by Tang et al. (2020) when dealing with irregularly gridded oceanic fields in the South China Sea.

3. Method (1) is a special case of "digital filters" (Shumann 1957; Feser and von Storch 2005)

$$M(f)_i = \sum_{k \in B} \beta_k f_{i+k} \tag{6.3}$$

with the finite "support basis" $B$, which is a small neighbourhood of all points. The original field is replaced by a field consisting of a weighted mean of close neighbours. If the overall mean value across $B$ is to be conserved, as in the case of large scales, then $\sum_k \beta_k = 1$. If only small scales shall be retained, then

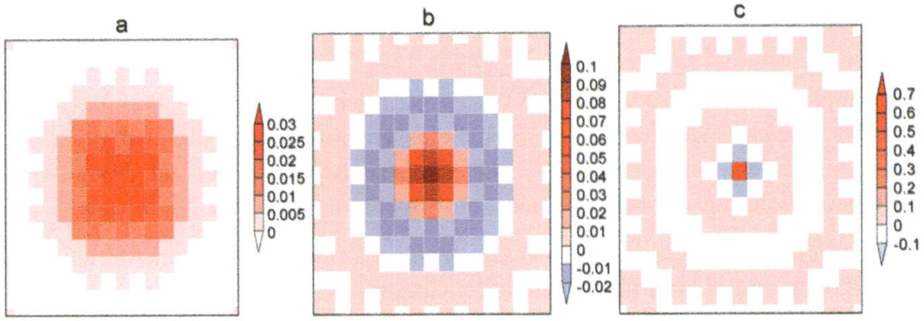


**Fig. 6.2** Limited support-bases $B$—with the point $i$ in the centre being replaced by the weighted average of all points in $B$, with weights as given by the colours. Here, 8 points to the left and right, to the top, and to the bottom, are chosen. (**a**) low pass, retaining only large scales, (**b**) medium pass, retaining medium scales, and (**c**) high-pass, retaining only small scales. For the response, see next Fig. 6.3 (Feser and von Storch 2005). © American Meteorological Society. Used with permission

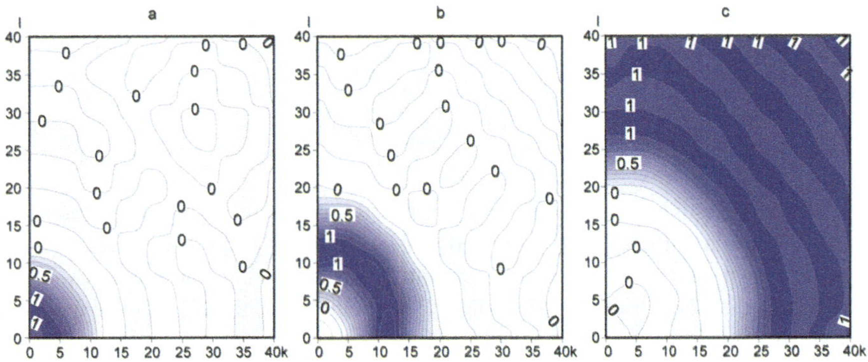


**Fig. 6.3** Filter response for the three filters shown in Fig. 6.2. The blue ares represent wavenumbers, which are maintained by the filter, while the wavenumbers in the white areas are mostly taken out (Feser and von Storch 2005). © American Meteorological Society. Used with permission

$\sum_k \beta_k = 0$. For a discussion of the benefits and drawbacks of digital filters, compared to Fourier filters, see Feser and von Storch (2005).

An example of three 2-dimensional filters, low-, band, and high-pass, is shown in Fig. 6.2, and the response, in the wavenumber space, in Fig. 6.3. The low-pass filter, in the left, removes most variability for wavenumbers beyond 10 (relative to the model domain, with the middle-range filter, only wavenumbers between about 5 and 20 are kept, while the low-pass-filter suppresses all variability related to wavenumbers < 20. An example of applying these filters is shown in Fig. 6.4). The scale-selective effect of the filters is obvious.

Note that such filters, apart from the EOF-based ones, can be used only for latitude/longitude or rectangular gridded fields such as those shown in Fig. 6.5.

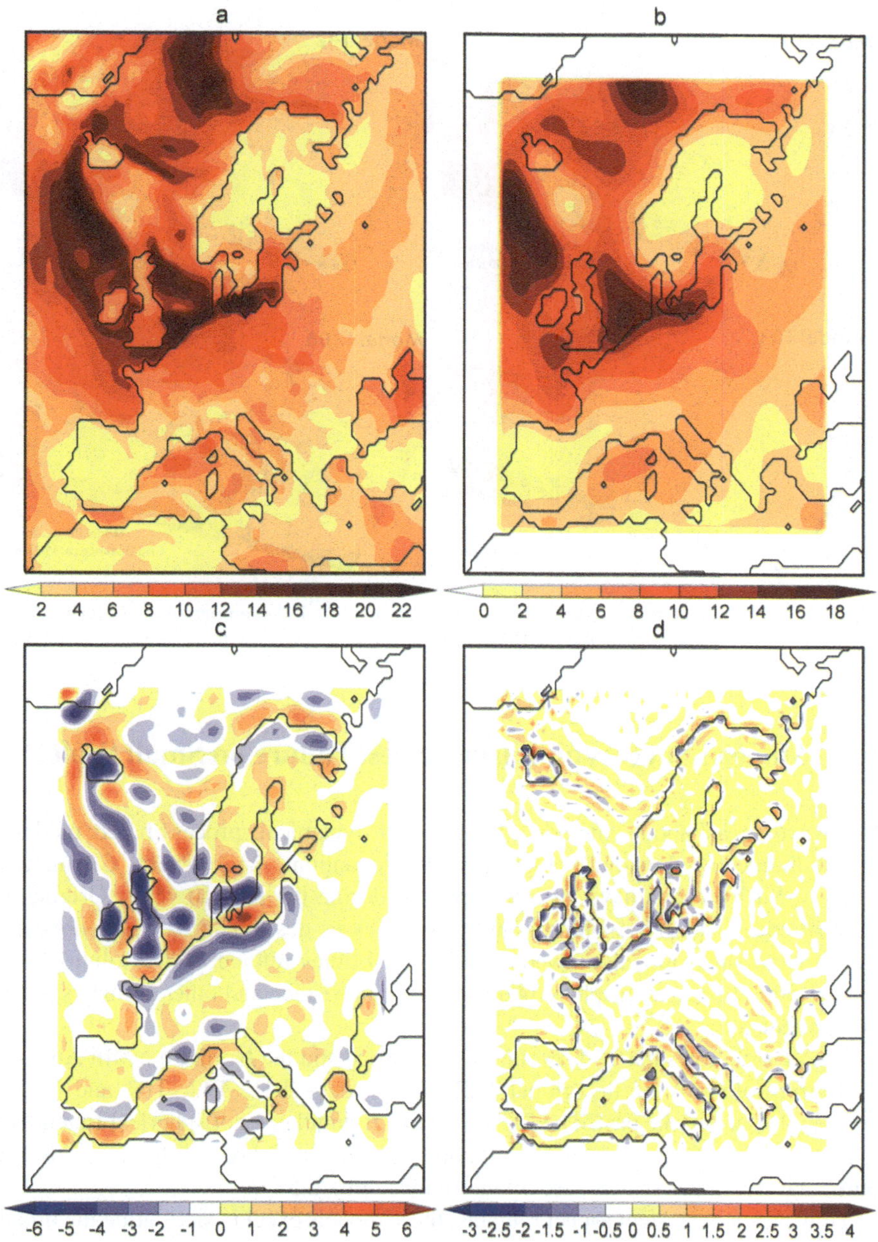


**Fig. 6.4** Example of the effect of the filters with filter weights (Fig. 6.2) and response functions (Fig. 6.3). The full field is given in (**a**), the loss-pass filtered fields in (**b**), the mid-range-filtered in (**b**) and the high-pass filtered in (**c**) (Feser and von Storch 2005). © American Meteorological Society. Used with permission

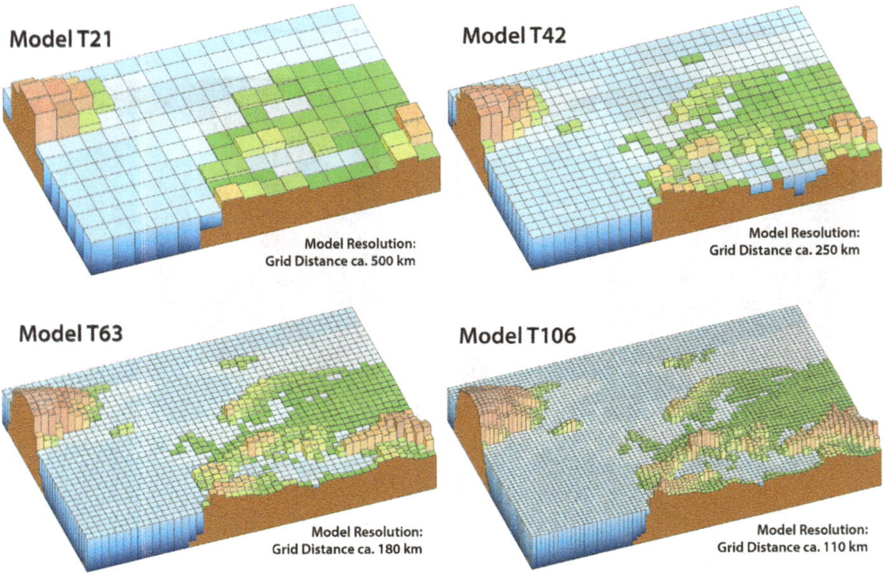


**Fig. 6.5** Conventional grid presentations "T21" to "T106" of the topography of Europe - as routinely used as standard in the ECHAM models until recently; with permission © DKRZ, Michael Böttinger

## 6.2 Models: Success on Large Scales, Failure on Small Scales

When we speak about "models", we refer to dynamical, process-based models, which mimic the dynamics of the primitive equations, i.e., General Circulation Models. Such models have greatly improved during the past 40 years. Originally of very coarse spatial resolution, and in most cases describing basically only the troposphere, such models nowadays have a 10-times refined grid resolution and feature not only the atmosphere but also an ocean, sea ice and other climate system components.

Figure 6.5 shows the grid resolution for Europe and the Northern North Atlantic of four resolutions of the global ECHAM model—T21, T42, T63 and T106, commonly used in the past. The "T" refers to a triangular set of spherical functions, which are transformed to the grid point space. Nowadays, different grids, in particular triangular set-ups, are often used (see Fig. 6.5).

However, a consistent aspect remains: these models excel in simulating the broad state of the climate system on a large scale. The fidelity of simulating specific features heavily depends on the chosen grid's resolution. At the finest spatial resolutions, where features are resolved by just one or a few grid boxes, climate models exhibit limited to no proficiency (Grotch and MacCracken 1991; Pielke 1991). Notably, when dealing with fields that exhibit substantial variation from one grid box to another, as frequently observed in precipitation patterns, the information

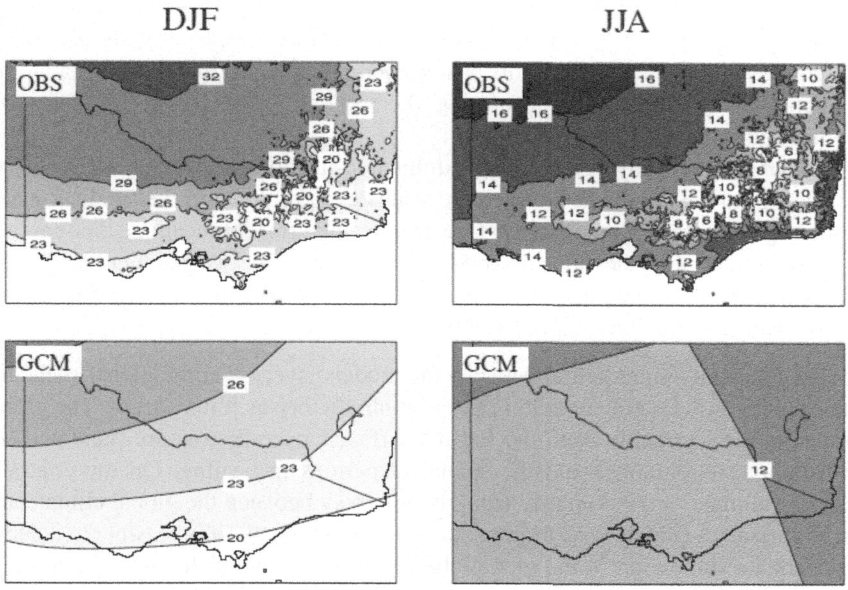


**Fig. 6.6** Seasonal mean precipitation amounts in State of Victoria in Southeastern Australia, in southern winter (JJA) and summer (DJF), as constructed from observations (top) versus simulations by a global model (Whetton et al. 2001) Creative Commons by Attribution (CC-BY) 4.0

obtained at the grid-box level is not reliable (Cubasch et al. 1992; von Storch et al. 1993; Whetton et al. 2001).

An example from these *old days* is provided in Fig. 6.6—maps derived from local observations of seasonal amounts of precipitation, and the same as simulated by a global climate model.[3] The diagram could be slightly misleading in the context of GCM output, as it presents a highly detailed map of the coastline that the GCM, by its nature, is unaware of. In reality, seasonal precipitation amounts exhibit substantial spatial variation, while the model implies a lack of discernible structure. During austral summer (DJF), however, the model does manage to replicate a basic land-sea pattern, with more precipitation inland and less along the coast.

Several factors contribute to the lack of success in modelling at the regional scale, which we define here as encompassing various mesh sizes within the model's grid. This implies that in a "T21" model, the 1000 km scale is considered regional, whereas in a "Tl06" model, 1000 km would be associated with large scales:

- The spatial resolution inadequately represents the Earth's surface structure. The distribution of land and sea is heavily blurred. In the late 1980s and early 1990s, most climate models operated at a "T21" resolution, which has now been

[3] Unfortunately, the grid resolution of the global model was not given, only a broad reference that such models would usually have grid lengths of 300–500 km.

replaced by the "T106" and better resolution (see spatial resolution of Europe in Fig. 6.5). In the T21 representation, mountains appear as wide and much reduced in height. Although this is improved in T106, the digital representations significantly fall short of capturing the steepness and height of real-world mountain ranges.
- The atmospheric hydrodynamics exhibit non-linearities, and the energy introduced into the system at the cyclonic scale cascades down to the smallest scales through nonlinear interactions. Due to the numerical truncation, this cascade is disrupted, and the flow towards the smallest scales is parameterised. This parameterization has the most pronounced impact on the smallest resolved scales. (see Roeckner and von Storch 1980).

The apparent contradiction between the models' success at the global scale and their failure at the regional scale is not as contradictory as it may seem. The global climate emerges mainly from the Earth's surface's overall structure (land and sea distribution, topography) and the spatial disparities in heating. On the contrary, regional climates arise primarily from the interplay between the global climate and local intricacies (the downscaling relationship). As a result, it is feasible to achieve an adequate simulation of the global climate even if none of the regional climates are accurately represented in their specific nuances.

**To put it differently, contrary to the perspective in the era of Vladimir Köppen, the global climate should not be seen as the mere sum of all regional climates. Instead, envision the global climate as the originator, or "mother", of numerous partially dependent "daughters," which are the regional climates.**

**In summary, climate models exhibit constrained spatial resolution. Features that are apparent over an area covered by numerous grid boxes are well-simulated, provided the underlying physics is accurately described. However, features confined to only a few grid boxes demonstrate lower fidelity in the simulation. Naturally, features smaller than the grid resolution are entirely absent from the depiction. Hence, the outcomes generated by quasi-realistic climate models encompass three distinct scale ranges. The first involves well-resolved scales (depicted in yellow in Fig. 6.7), followed by scales that receive inadequate resolution (indicated by red hatching), and finally, scales that are entirely overlooked. Downscaling comprises an ensemble of techniques aimed at establishing dependable statistical information within the inadequately resolved and unresolved domains. In the subsequent section, we will elaborate on the principles of downscaling methods and underscore their potential through illustrative examples.**

A recent development by the German Weather Service and by the Max-Planck Institute of Meteorology in Hamburg is the model ICON, which operates on a grid of triangles. *In future*, climate simulations may be done regularly with the "R2B8"-triangle resolution, corresponding to a grid resolution of about 10 km, and process studies will resolve large clouds in "R2B11", corresponding to a 1.25 km grid (Fig. 6.8). This does not change the different scale-dependent performance of such models, but separating scales is still a challenge: The methods shown before in

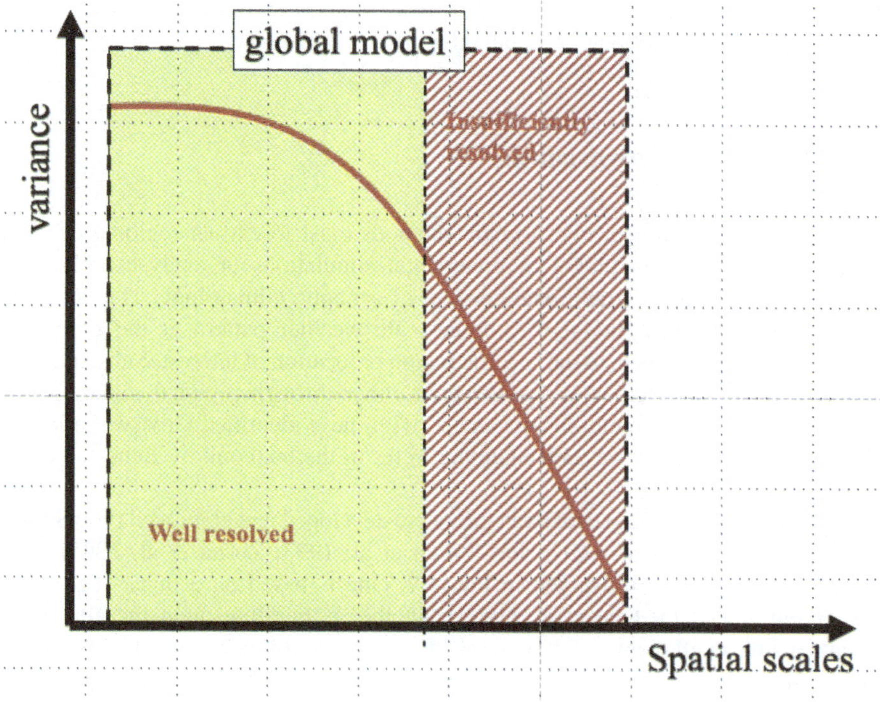


**Fig. 6.7** Scale ranges of the output of a global atmospheric model: Large scales to the left, small scales to the right. The plotted curve illustrates the energy distribution or variance across these scales. The segment highlighted in yellow signifies the scales at which the model's output is perceived as realistic, while the hatched red region represents compromised dynamics due to inadequate resolution. The exact location of the boundary between the yellow and red regions remains uncertain, although it has been occasionally hypothesised to be at a minimum of four grid lengths (Pielke 1991). From (von Storch and Zorita 2019) CC BY

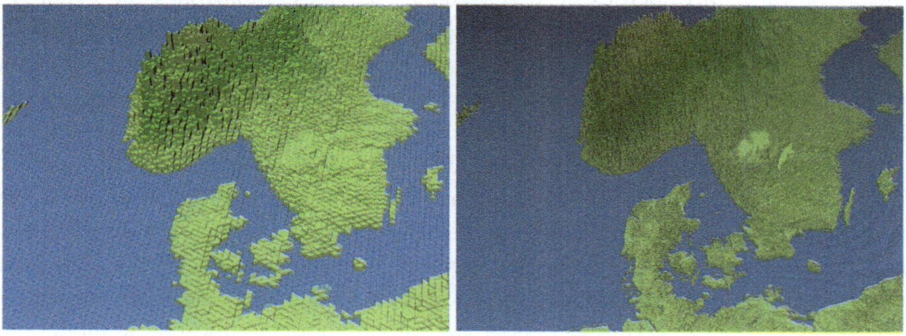

**Fig. 6.8** ICON grids R2B8 and R2B12 covering Southern Scandinavia; with permission © DKRZ, Michael Böttinger

Sect. 6.1 cannot be used for triangular "ICON"-fields (Fig. 6.8), except if the fields are interpolated to a lat/long or rectangular grid.

## 6.3 Empirical Downscaling

As previously mentioned, two primary methods exist to extract regional or even local information from global or continental simulations or analyses. The first method is rooted in the traditional concept of synoptic climatology. This involves identifying a "Grosswetterlage", a large-scale weather pattern at the global or continental level pertinent to the specific region or location of interest. Subsequently, coherent regional specifics are constructed either through regional modelling or by studying regional conditions associated with these identified Grosswetterlagen. For a more comprehensive understanding, refer to the textbook by Benestad et al. (2008).

In Hamburg, three statistical strategies were developed and employed (von Storch et al. 1991; Zorita et al. 1992; von Storch et al. 1993; Zorita et al. 1995)—the underlying concept is illustrated in Fig. 6.9. One is based on a linear regression earlier introduced by Kim et al. (1984), another is building upon the analogue-ansatz (using coefficients of large-scale EOFs as coordinates, and local rainfall on the Iberian Peninsula to be interpolated in this space). Both are still widely influential. Later, another method, from the tradition of geostatistics, was developed and tested by Biau et al. (1999), again using EOF coefficients as coordinates used for interpolation. The different methods were compared by Zorita and von Storch (1999).

The skill of the regression-based (mostly using canonical correlation) and of the analogue methods turned out to be similar (cf. Fig. 6.11).

These methodologies adeptly establish statistical connections between large-scale states and small-scale state variables, encompassing measures such as time means or variability metrics. The subsequent section demonstrates the extensive spectrum of possibilities through three examples utilizing the regression method.

The process of downscaling (see Fig. 6.9) commences with the identification of the "predictand", a regional parameter denoted as $\mathcal{R}$, which is intended to be estimated conditional upon a set of given large scale "predictor" $\mathcal{L}$. This parameter, or variable, can take on diverse forms. For instance, it might represent monthly average regional rainfall (refer to Sect. 6.3.1), or the seasonal variability in storm surge heights (see Sect. 6.3.2), or even the parameters of a stochastic rainfall generator (covered in Sect. 6.3.3). The first example could also be effectively addressed using an atmospheric limited area model (thus involving dynamical downscaling). The second instance would necessitate not only a regional atmospheric model but also a coastal sea dynamical model. In the third case, a stochastic model would need to be fitted to the outcomes of a regional dynamical model. However, there are cases where empirical downscaling is the sole viable approach, such as in the

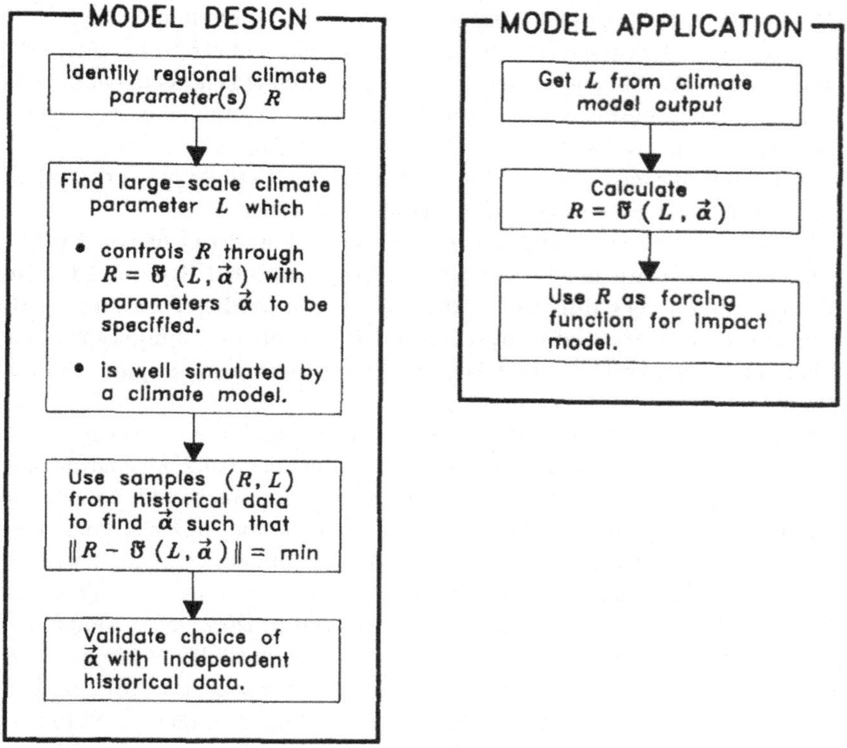


**Fig. 6.9** Sketch of the concept of empirical downscaling. After von Storch (1995) CC BY

instance of plant flowering, exemplified by the plant Galanthus Nivalis in Germany's Schleswig-Holstein as discussed by Maak and von Storch (1997).

The subsequent phase involves the identification of a predictor $\mathcal{L}$, which not only exhibits a robust connection with $\mathcal{R}$, but is also consistently furnished by simulations of climate change scenarios and by historical analyses. Typically, these predictors encompass attributes that exhibit gradual shifts rather than sudden fluctuations, including time mean temperature and air pressure.

The link between $\mathcal{L}$ and $\mathcal{R}$ is forged through an empirical model $\mathcal{F}$, which details the relationship between $\mathcal{L}$ and $\mathcal{R}$ provided appropriate parameters $\alpha$ can be determined. These parameters $\alpha$ are determined by fitting $\mathcal{F}(\mathcal{L}, \alpha)$ to $\mathcal{R}$ for a large set of pairs of $\mathcal{R}$'s and $\mathcal{L}$'s. If more data are available, these may be used to test the empirical model with independent data.

On the right side of Fig. 6.9, the application of the previously determined empirical model is sketched, which is straightforward. First, the variables $\mathcal{L}$ from an analysis or a simulation are extracted and then processed with $\mathcal{F}(\mathcal{L}, \alpha)$, The resulting estimated $\mathcal{R}$ may then be used in impact studies.

When the simulation is a projection of future climate change, there is an important caveat—namely the assumption that the found link between the large scale variable and the regional variable is unchanged in this future climate.[4] For one configuration, this assumption was tested and found valid (Busuioc et al. 1999).

When using the analogue-method, no fitting or training is needed, but the space to determine the nearest neighbour, and the measure to determine distances are left to be chosen by the researcher. In the case of Zorita et al. (1992), the space was chosen as being spanned by the first few large-scale EOFs of air pressure. Different choices of measures were examined by Matulla et al. (2008).

**The "trick" with empirical downscaling is that statistics of two variables are linked, which are not necessarily directly connected dynamically. They typically vary on very different spatial scales: the predictor on large scales, which presumably are well represented in climate change simulations, and a predictand on small scales, which is needed for estimating an impact of climate change.**

The statistical strategy presents various advantages and limitations compared to the nested limited-area model approach discussed in Sect. 6.4.2. One advantage is its technical simplicity. Another advantage lies in its alignment with observations, often lacking in the GCMs themselves, thereby enforcing physical consistency. Additionally, this approach is not confined solely to physical regional parameters like rainfall or temperature but can also be extended to economic (e.g., agricultural yield) or ecological parameters (e.g., tree ring width). However, a limitation of the statistical approach arises when there is not a strong relationship between a large-scale parameter and regional climate, which is often the case. Also, like many statistical tools, its validity might not extend beyond the data range used to fit the model, assuming a stationary relationship between anomalies.

### *6.3.1 The Classical Case: Iberian Precipitation*

Zorita et al. (1992) determined a statistical model which relates the irregularly distributed Iberian Peninsula (Spain and Portugal) winter mean rainfall field to the gridded observed winter mean sea level pressure field in the Atlantic Sector, using data from 1950–1980. To do so, with the help of a canonical correlation analysis (CCA) a link was constructed between simultaneous variations of the two fields; one dominant pair of patterns was found (Fig. 6.10).

The pair of patterns represents a reasonable physical relationship: if the anomalous[5] sea level pressure pattern $\mathcal{L}$ has a positive coefficient, then the mean advection of maritime air from the ocean to the Iberian Peninsula is intensified. The process of lifting humid air at the coastal mountains of the Peninsula is more efficient than

[4] Such an assumption is also postulated with respect to the parameterizations operating in dynamical climate models when used to study uncharted future climates.

[5] i.e., the deviation form the long term mean.

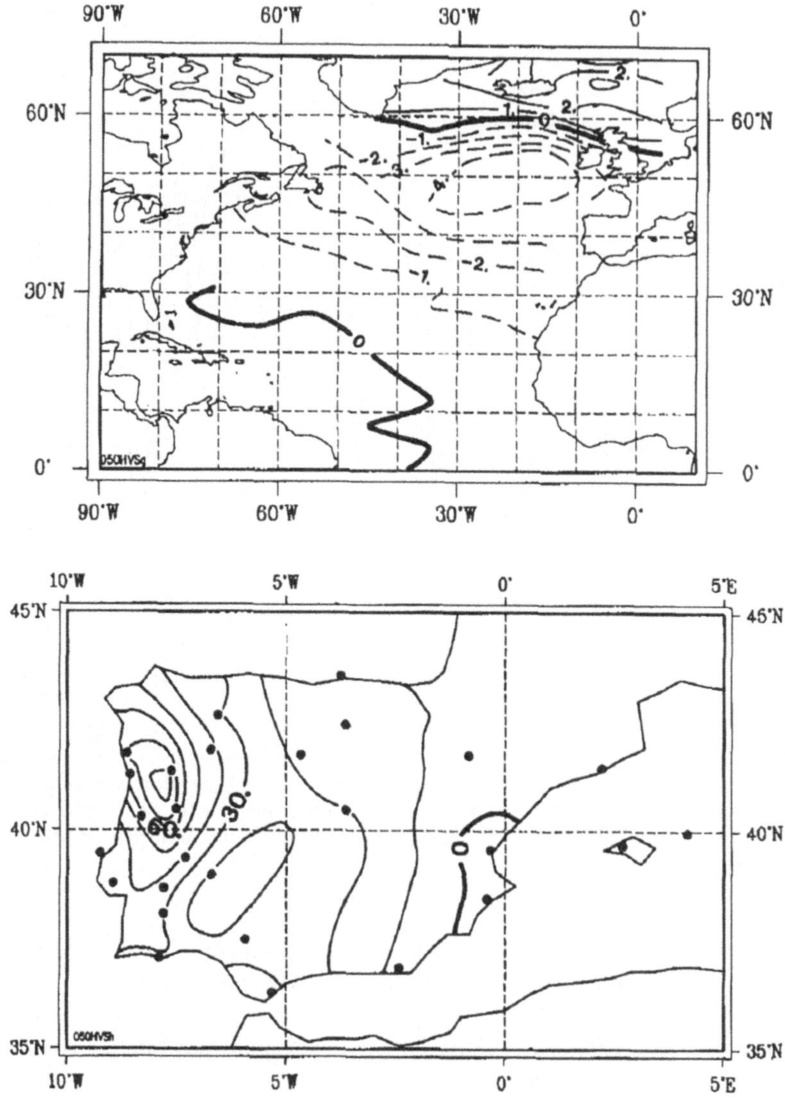


**Fig. 6.10** Pair of patterns, determined with Canonical Correlation Analysis, which go with maximum correlation, which is 0.75. Top: winter mean air pressure anomaly. Bottom: winter mean precipitation anomaly. These patterns explain about 65% and 40% of the respective total variances. The SLP data (mb) are on a $10° \times 4°$ lat/long grid available. The rainfall data (mm/month) comprise 29 stations, the locations of which are marked by dots. From Zorita et al. (1992) Published 1992 by the American Meteorological Society

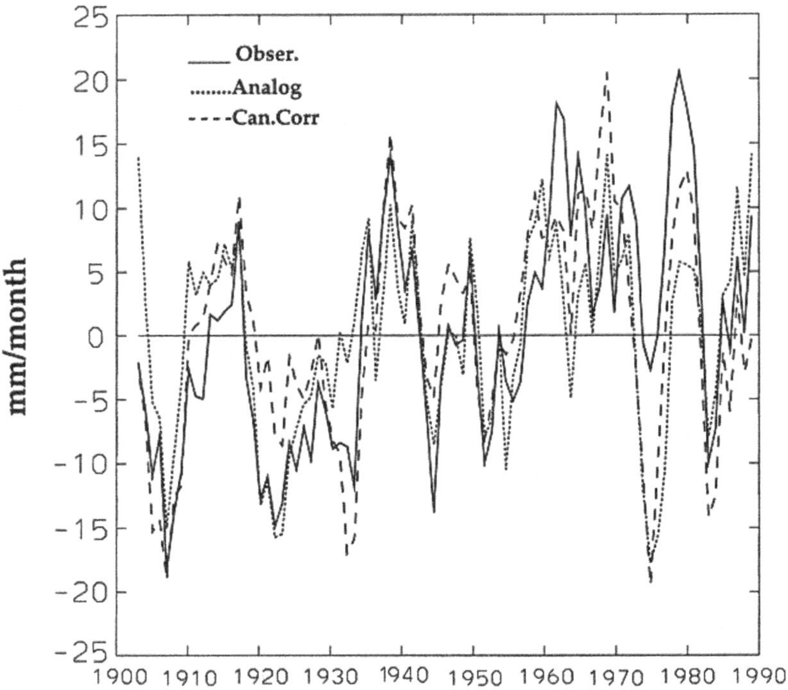


**Fig. 6.11** Five-year-running mean time series of area-averaged DJF rainfall anomalies (mm/month) as derived from station data on the Iberian Peninsula (solid line); derived from the North Atlantic SLP pressure field: analog method (dotted line) and canonical correlation analysis (CCA, dashed line). The data 1951–1989 are used for fitting, the data for 1901–1950 are used to validate the approach. Published 1992 by the American Meteorological Society (Zorita and von Storch 1999)

on time average, and rainfall $\mathcal{R}$ is enhanced. If the coefficient is negative, the mean advection is weakened and, consistently, less rainfall is happening.

This empirical model was used to build a first empirical downscaling model (von Storch et al. 1993). This case became quite popular, some may think: a classical, as documented by many citations.

First, the empirical model was tested if, in an additional time period, namely 1901–1950, the model would reproduce the observed small-scale variable, i.e., seasonal mean precipitation (Fig. 6.11), and then projections of the possible future state of the large-scale driver, i.e. sea level pressure, were downscaled to obtain plausible future seasonal mean precipitation.

Simulation data generated in a pair of two 100-year simulations with the then state-of-the-art climate (atmospheric horizontal gridding: T21) model of the Max-Planck Institute for Meteorology (Cubasch et al. 1992) was examined, one with present day $CO_2$-concentrations, and another with doubled $CO_2$ concentrations. The long-term winter-mean precipitation change at the four grid points covering the Iberian Peninsula was negative (details see Fig. 6.12b), with a mean decrease of about −10 mm/month. When using the downscaling methodology, the estimate using the change in North Atlantic SLP resulted in a weak but spatially detailed increase in precipitation (Fig. 6.12a; 1 mm/month), with maximum values of 4 mm/month on the western side of the peninsula.

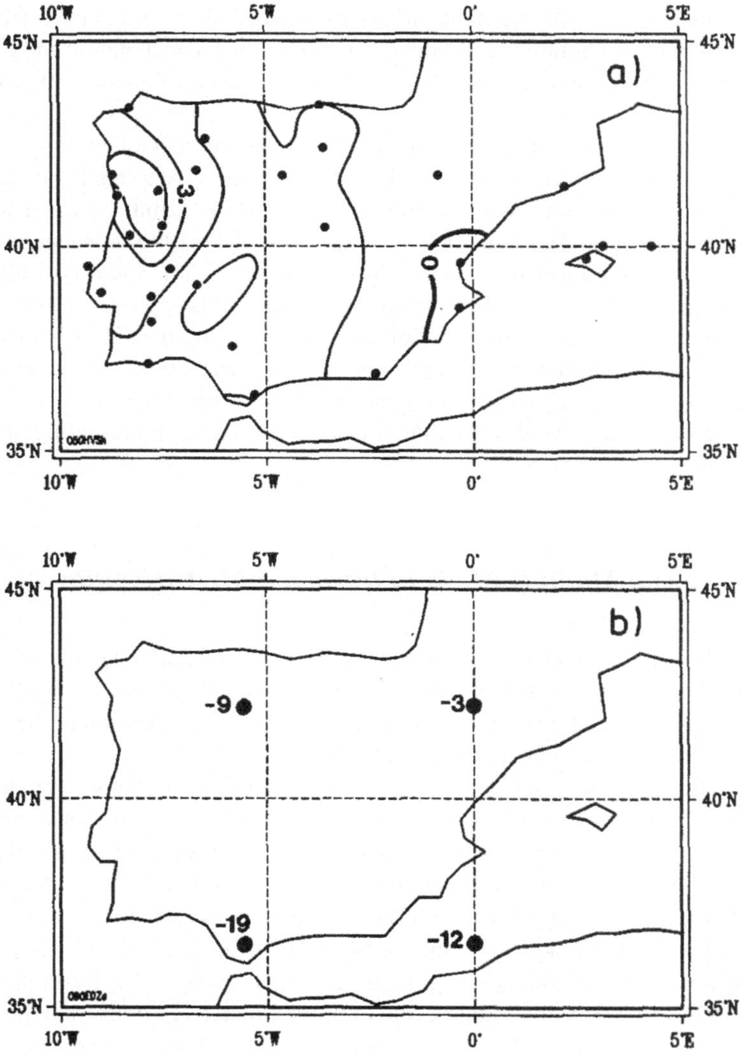


**Fig. 6.12** Regional change of winter Iberian rainfall (mm/month) in the "2 $CO_2$" experiment. (**a**) Indirectly derived from the simulated change of North Atlantic SLP field; and (**b**) directly simulated by the GCM. © American Meteorological Society. Used with permission (von Storch et al. 1993)

The discrepancy in both the sign and pattern of the directly simulated and estimated rainfall changes suggests that the mechanisms influencing Iberian rainfall in the GCM differ from those observed in reality. This once more underscores the GCM's lack of precision at regional scales.

von Storch et al. (1993) argued that the statistical downscaling method effectively translates comprehensive information from climate model simulations that is believed to be reliable—into regional scales that are pertinent for specific users, such as hydrologists. In forecasting terms, the method aligns with the "perfect prog" approach, wherein a statistical model is established between dynamic and predictive quantities in the observed atmosphere and then applied unchanged to the simulated atmosphere.

The empirical downscaling methodology is reminiscent of the approaches of Karl et al. (1990) and of Wigley et al. (1990), but there is a significant difference with respect to the scales. Both approaches establish a link between local and regional-scale variables. Adhering to Wigley et al.'s (1990)-concept, a suitable estimator for local rainfall anomalies in the Iberian region would typically be the anomalous spatially averaged precipitation. However, this approach falls short in our case because the area-average precipitation simulated by the GCM is inconsistent with the simulated changes in the North Atlantic circulation—the genuine factor influencing winter Iberian precipitation. To circumvent this challenge, mean sea level pressure enters the downscaling method as predictor, representing large-scale circulation.

### *6.3.2 From Monthly Mean to Intramonthly Variability*

While the case presented in the previous Sect. 6.3.1 is straightforward, we now revisit a less direct case, which relates the large-scale monthly mean circulation over the Northwest Atlantic and the statistics of high tides in Cuxhaven at the mouth of the river Elbe (von Storch and Reichardt 1997).

Variations of the high tides are mostly related to varying regional winds associated with storms, while the time mean circulation is, by itself, irrelevant. However, anomalies of the mean circulation have been found to be associated with changing storm tracks and intensities (e.g., Branstator 1995). Thus, it is plausible that changes in circulation carry information about changing storm surges.

The predictand is derived from the monthly statistics of storm surges at Cuxhaven, observational data covering high tide water levels at that location spanning from 1876 to 1993. With a lunar cycle of approximately 12 hours and 25 minutes, there are roughly two observations per day. This chronological data captures diverse influences, including the impact of storm-induced winds, sea level rise, land subsidence, alterations in tidal dynamics due to human management along the coastline and changes in the Elbe River's bottom terrain. The annual mean sea level shows a notable upward trajectory, with the high tide's average level in 1990 standing approximately 30 cm higher than at the start of the recordings in 1876. For deriving only fluctuations in storm-related water levels the annual means of water levels are subtracted. This process helps prevent potential gradual irregularities that could arise from various factors, like trends linked to rising sea levels or human activities such as harbour dredging.

For each month, three percentiles for the high tide variations were calculated, the 50%, the 80% an the 90%. The 90%-ile is larger than about 54 of recorded daily water levels, the 80%-tile larger than 48 daily water levels. The vector of these three percentiles is the predictand. These are measures of the intra-monthly variability.

The data to deal with the predictor dataset involves monthly evaluations of large-scale air pressure distribution across western Europe and the northern North Atlantic (30°–7°N, 70°W–25°E) from 1899 to 1988. While the daily assessments may contain inconsistencies, the monthly evaluations are considered mostly consistent and not significantly affected by irregularities (Trenberth and Paolino 1980).

The predictand and the predictor are linked through a Canonical Correlation Analysis, which determines patterns in both, the predictand and the predictor, that are most strongly linearly related, using the data 1970–1988. The remaining data, prior to 1970, are kept as independent data to verify the statistical model. For simplicity, only the first pair of patterns is discussed in the following.

The first, and most strongly linked pair, is made up of the pattern shown in Fig. 6.13 and of the three component vector (−21, −16, −18 cm): with −21 cm for the 50%ile, −16 cm the 80%-ile and −18 cm the 90%ile. This link explains on average 83% of the variance of the percentiles, and the correlation amounts to 0.89. Thus, a monthly mean flow anomaly, as shown in Fig. 6.13 is associated with a reduction of storm surge heights, for instance for the 90%-ile by −18 cm. Since the Canonical Correlation analysis is a linear analysis, the same holds with reversed sign. The link makes physical sense—if we have an anomalous low-pressure system southeast of the British Islands and a high-pressure system over Northern Scandinavia, a southeasterly flow passes over the North Sea. Normally,

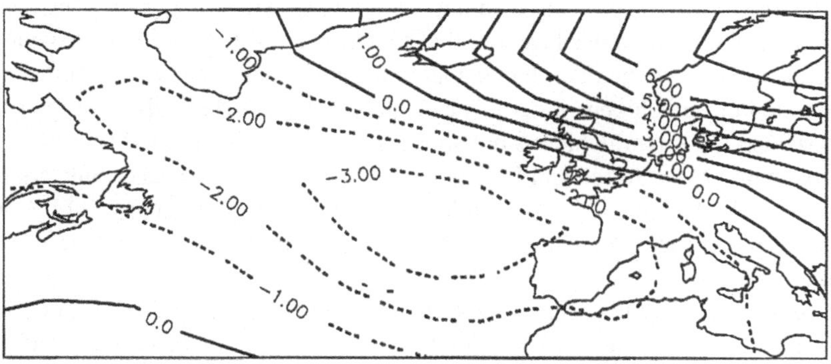


**Fig. 6.13** Regional change of North Atlantic sea level pressure anomalies (hPa) with the strongest link to the percentiles of local high tide variations in Cuxhaven (at the SE "corner" of the North Sea). Published 1997 by the American Meteorological Society (von Storch and Reichardt 1997)

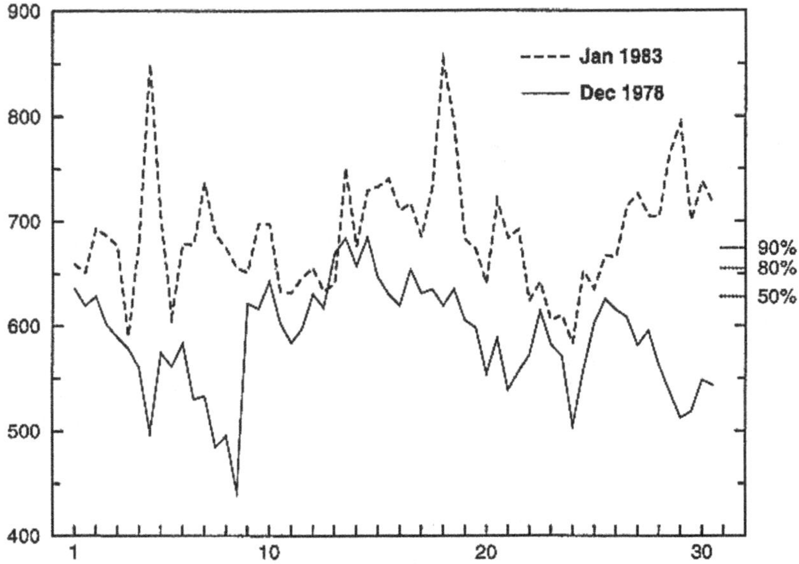


**Fig. 6.14** Time series of high-tide levels monitored at Cuxhaven during January 1983 and December 1978 with a large positive, or a negative, CCA coefficient. Vertical axes: water levels in centimetres. Horizontal axes: time in days. The climatological mean values of the 50%, 80%, and 90% percentiles are given as tick marks on the left margin of the diagrams. Published 1997 by the American Meteorological Society (von Storch and Reichardt 1997)

there is a westerly flow over the North Sea, carrying several storms eastward; thus, when this pattern prevails, fewer eastward, surge-relevant storms prevail.

Figure 6.14 shows the intra-monthly variations of sea level at Cuxhaven for two months, when the coefficient of the first CCA pair is high or low, namely in January 1883 (dashed), or December 1978 (continuous line). According to the link, a large positive coefficient would lower all three percentiles, indicating that the whole distribution of water levels in that month is reduced—and that is obviously so in this month. On the other hand, a large negative coefficient will shift the entire distribution towards higher water levels, which is also clearly visible in Fig. 6.14.

A second CCA pair, which is not shown here because of limited space but may be looked up in von Storch and Reichardt (1997), describes a weaker link, which is less obviously dynamically explainable. It describes almost no change of the 50%-ile, but a clearly increased higher percentiles, which alludes to a single stronger storm.

When interpreting the scenario with a doubled $CO_2$ concentration, as described in Sect. 6.3.1, a change of 7 cm, 6 cm, and 7 cm for the 50%, 80%, and 90% percentiles is suggested for the time of doubling atmospheric $CO_2$ concentrations—on top of the rise in mean sea level. These are "expected values", given by the specific pair of simulations; using a different pair of simulations will result in different estimates of the change; also, because of the statistical nature of downscaling link, the changes may also deviate to some extent form these numbers. Since the indicated numbers are rather small, one may summarise the projected change as "small", in particular when compared to the rise in mean sea level.

### 6.3.3 *From Monthly Means to Parameters of a Rainfall Generator*

The third case also deals with an indirect link, namely using monthly mean maps of air pressure on the continental scale, which condition the parameters of a rainfall generator for a location, here: Bucharest in Romania (Busuioc and von Storch 2003).

The rainfall generator is given by a 2-state, first-order Markov chain.[6] Precipitation either occurs or not (the two states), and the conditional probability of precipitation occurrence depends only on the occurrence on the previous day. Two parameters describe the precipitation occurrence process: the transition probabilities $p_{01}$, of a wet day following a dry day, and $p_{11}$, the probability of a wet day following a wet day. A daily precipitation amount of >0.1 mm is adopted as a wet-day threshold.

The variation of precipitation amount on wet days is sampled from a gamma distribution with a shape parameter ($k$) and the scale parameter ($\beta$). The mean precipitation amount on wet days is $\mu = k\beta$. From a time series of daily precipitation amounts, the four parameters may be derived. For Bucharest, such daily precipitation amounts exist since 1901.

The downscaling link relates maps of European (5–50° E, 30–55°N) SLP seasonal means and the vector of the four parameters ($p_0$, $p_{11}$, $\mu$, $k$) with the help of a CCA.

The full data set, 1901–1999, was split into two intervals, 1901–1949 and 1950–1999. One of the intervals was used to fit the CCA, and the other was used to determine the skill of the empirical model in generating a realistic series of rainfall for Bucharest. The case shown below uses the data 1901–1949 to fit the data and the interval 1950–1999 to determine the skill.

The CCA identifies pairs of patterns in the SLP fields and in combined vectors of the transition probabilities and gamma distribution parameters whose time series share a maximum of correlation. The identified patterns are strongest in winter and autumn. For the sake of brevity, only winter results are shown and only the first CCA pair (even if the second is also carrying significant information).

Figure 6.15 shows this first CCA pair in winter. The associated time series are normalised to exhibit a standard deviation of 1; as usual with CCA, the signs of the patterns, both together, may be reversed. Southerly/northwesterly circulation over Romania in winter is associated with above/below-normal daily mean precipitation

[6] The following is in part quoted verbatim from Busuioc and von Storch (2003).

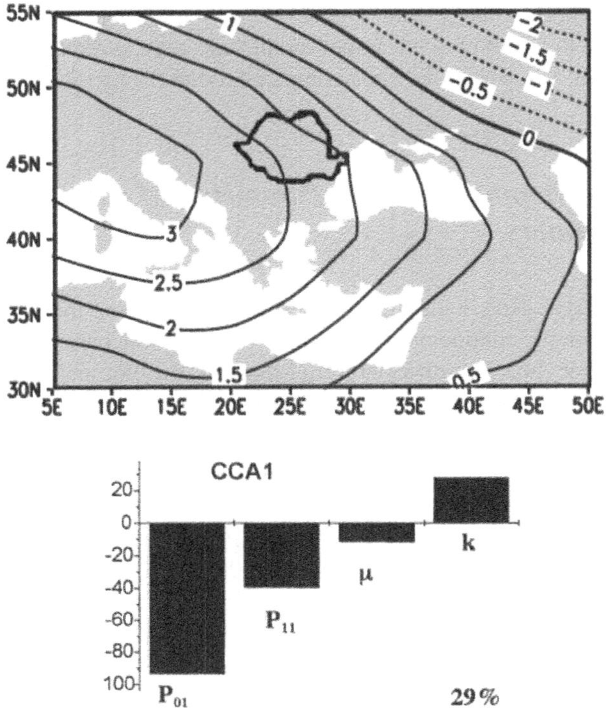


**Fig. 6.15** Patterns of the first CCA pair of winter SLP anomalies and of the winter anomalies of rainfall generator parameters derived from the first half of the observations (1901–1949). The SLP pair represents 77% of the 1950–1999-variance of SLP variance, the parameter-vector 29% (Busuioc and von Storch 2003). CC BY

on wet days and higher/lower transition probabilities to wet days at Bucharest. Southwesterly circulation brings moist air masses from the Mediterranean basin to Romania (especially in the southern part, where Bucharest is located), and more precipitation is recorded. As a result, the daily mean precipitation $\mu$ on wet days and the wet day probability ($p_{01} + p_{11}$) are higher. The seasonal total is dominated by the number of wet days and is less dependent on the mean amount of wet days.

Figure 6.16 shows the time series of the transition probabilities, as determined from the local data (solid line) and through the downscaling model (dashed line), which was fitted to the 1901–1999 data. Obviously, the empirical model recovers most of the variability of both transition probabilities.

The skill (in terms of correlations and explained variance) of the statistical downscaling model, derived from the 1901–1949 fit), using the time coefficients in the independent time interval 1950–1999 finds the model is most skilful for the transition probabilities, slightly skilful for $\mu$ unskillful for $k$.

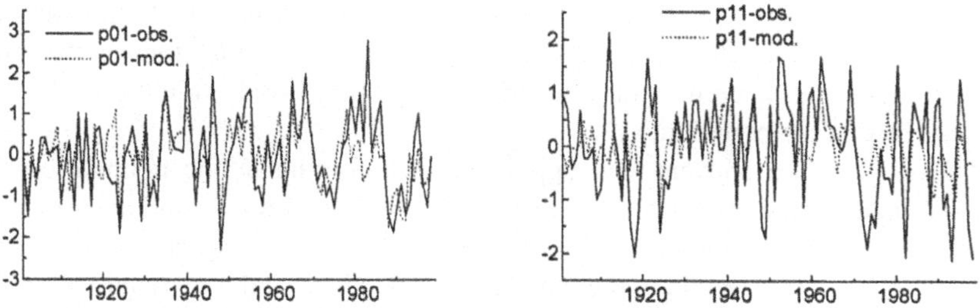


**Fig. 6.16** Winter standardized anomalies of the transition probabilities $p_{01}$ and $p_{11}$ for 1901–1999 derived from observations (solid line) and derived indirectly from the observed European-scale SLP anomalies using the downscaling model (dashed line) fitted to the 1901–1949 data (Busuioc and von Storch 2003). CC BY

In Busuioc and von Storch (2003), the downscaling system is examined with respect to its skill in describing a number of characteristics, which will not be detailed here for the sake of brevity:

1. Appearance/nonappearance of precipitation quantified by mean and expected maximum duration of wet and dry intervals.
2. Daily mean and standard deviation of precipitation for wet days, expected maximum daily precipitation amount and frequency distributions of daily precipitation.
3. Changes (linear trend) in the seasonal precipitation.
4. Interannual variability of seasonal precipitation amount quantified by the standard deviation.

## 6.4 Dynamical Downscaling

### *6.4.1 Conventional Limited Area Modelling*

The practice of modelling atmospheric dynamics on a regional area, first for forecasting and later for climate simulations, is done since many decades.[7] Pioneers in this field are Davies (1976) and Giorgi (1990). The benefit of doing so was that simulations with greatly enhanced spatial resolution could be done with the same computer time. Davies' trick was to add a "sponge zone" along the lateral boundaries, where the interior state was nudged towards the driving global simulation or analysis, avoiding the problems of different phase speeds in the forcing field and in the interior.

Originally, the method was developed for weather forecasting and not for long simulations—in that case, the evolution in the interior of the limited area is only

[7] For an overview, consider, e.g., Takayabu et al. (2016).

partly determined by the boundary values but also by the initial conditions; when used in the climate mode, when the system is integrated for much longer, the details of the initial values become irrelevant, and the trajectory is only conditioned by what is prescribed at the boundaries.

The skill of the conventional approach, with forcing at the boundaries, was demonstrated by the "Big-Brother"-experiment by René Laprise and his coworkers (Denis et al. 2002a). This experiment consists of two simulations with the same horizontal resolution, a global one named "Big Brother" and a regional one named "Little Brother". The latter runs on a regional subarea of the former, and is conditioned at its lateral boundaries with the state of the Big Brother, however, after deleting all small scale details; also the initial state is the same as in the global simulation, but also after coarsening. In the beginning, the humidity distribution of the regional model exhibits less spatial detail, but already two days later, the details of the Big Brother reappear in the Little Brother experiment (Fig. 6.17) in

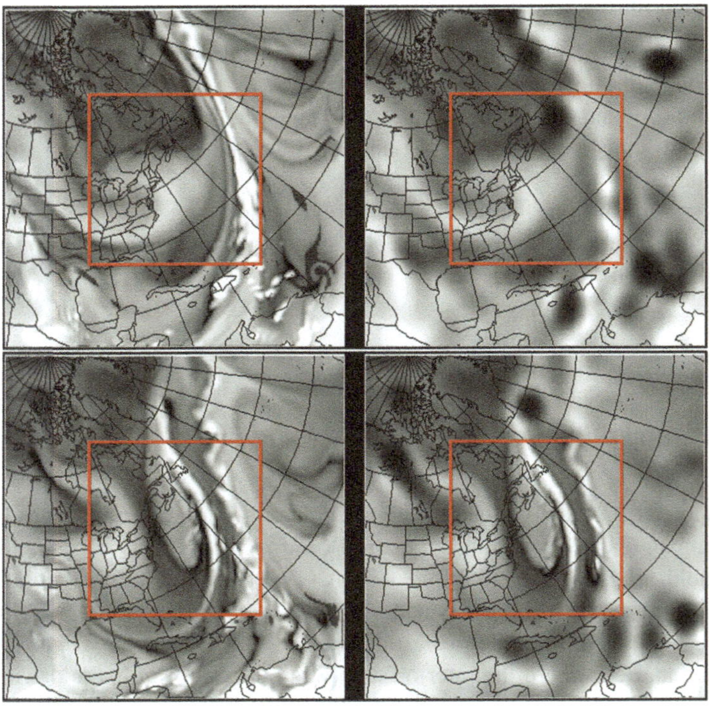

**Fig. 6.17** Specific humidity at 700 hPa on day 0 (top), and on day 2 (bottom). The left column is the control Big Brother simulation with the same grid resolution as the limited area model run in the red box. In the maps on the right side, outside of the red boxes, the coarsened fields from the Big Brother are shown, which provide the boundary values for the Little Brother. Within the inner red squares, the Little Brother simulation is done with the same fine resolution as the original Big Brother (Denis et al. 2002b)

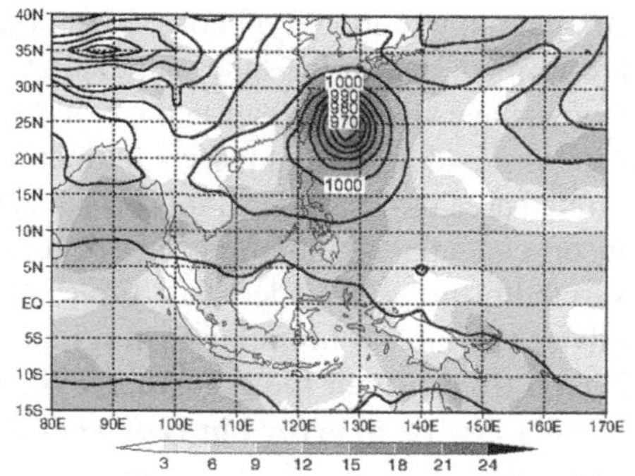


**Fig. 6.18** Air pressure ($hPa$; isobars) and wind speed analysis ($ms^{-1}$; shaded) of the typhoon Winnie at 0000 UTC 17 Aug 1997 in the NCEP–NCAR reanalysis. © American Meteorological Society. Used with permission (Feser and von Storch 2008)

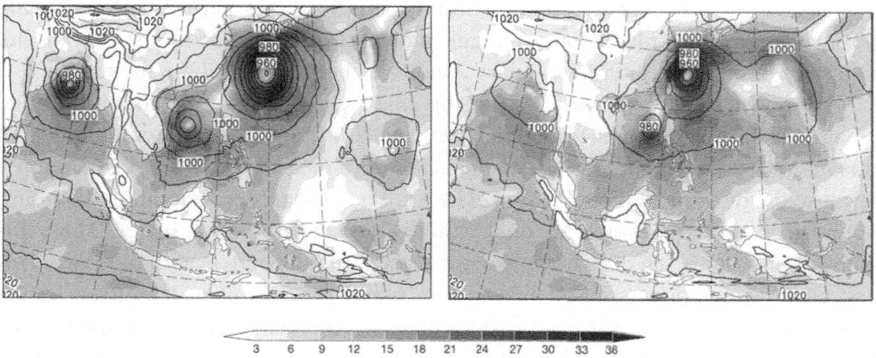


**Fig. 6.19** Simulation of air pressure distributions (hPa; isobars) and wind speed distributions ($ms^{-1}$; shaded) of the typhoon Winnie at 0000 UTC 17 Aug 1997 in two limited area simulations, using a grid of 50 km, begun before the formation of the storm, with (left) initial conditions of 3 Aug 1997 and (right) of 7 Aug 1997. © American Meteorological Society. Used with permission (Feser and von Storch 2008)

spite of the coarsening at the lateral boundaries.[8] **This experiment demonstrates that a regional model is indeed capable of recovering small-scale details from lower resolution boundaries**. Of course, the success of this experiment depends on several factors, namely on the strength of the stipulated west-east through flow, on the detail provided at the surface, and on the size of the regional model domain.

Another case of a "downscaling" with a limited area model refers to the case of the East Asian typhoon "Winnie" in 1997 (Feser and von Storch 2008). Its position and intensity are shown in Fig. 6.18 as given in the NCEP-NCAR reanalysis. The skill of reproducing this storm is analyzed in two extended simulations with a limited area model and a spatial grid resolution of about 50 km—in climate mode, i.e., with different initialisation times before the formation of the storm.

Both simulations (Fig. 6.19) capture the typhoon reasonably right, with a reasonable deepening compared to the lower resolution re-analysis. However, in one

[8] The effect of the coarsening can be seen on the right side, outside of the red box.

simulation, two more typhoons, in the Bay of Bengal and in the South China Sea, are formed, and in the other simulation one storm west of the Philippines. These storms have no counterpart in the real world analysis.

One could argue that these additional storms are an indication of an error in the limited area model. Indeed, it is probable that all model codes contain a few mostly inconsequential minor errors, but a better explanation is that the unavoidable internal variability in such models (see Chap. 2) is behind these divergent developments. **Given the lateral boundary values, such different developments are possible, even if they did not really form. Or is plausible that the differences to "reality" are not due to an error but represent the degree of uncertainty inherent in the dynamics of the weather system.**

While this seems to be in conflict with the Big Brother experiment, it is actually not—the region considered here is larger, and the information about the storm is not imported through the lateral boundary conditions but formed within the region.

The conventional approach of deriving small-scale regional detail with limited area models forced with lateral boundary values is often named "downscaling", but it is *not* projecting changes on a smaller scale from the state at larger scales, simply because the dynamical model is processing not large scales but lateral (and bottom) boundary conditions, which are only partially connected to large scales. Thus, this conventional approach does not make use of the scale-dependent skill of re-analyses and GCM simulations, as sketched in Fig. 6.7. The state on scales, given by the yellow part in Fig. 6.7, are supposedly "well resolved" and reliably described. Thus, the limited area model should use these large-scale states as drivers (plus the lateral conditions to close the numerical problem).

## 6.4.2 Spectral Nudging of LAMs

Thus, the proper approach is that of a state-space problem (e.g., Robinson et al. 1998), with a dynamical model and an observational equation:

$$\Psi_{t+1} = \mathcal{F}(\Psi_t, \eta) + \epsilon \tag{6.4}$$

$$\Omega_t = \mathcal{G}(\Psi_t) + \delta \tag{6.5}$$

Equation (6.4) is called a *state space equation*, as it describes the propagation of the state space variable $\Psi$ in time, influenced by some driving factors $\eta$. The term $\epsilon$ is an error term, added in recognition that the propagation is just an approximation. $\mathcal{F}$ is a dynamical model such as a limited area model.

Equation (6.5) is a *observation equation* which relates certain observable $\Omega$ to the state variable $\Psi$. Also, this equation is often not perfect but sometimes suffers from inaccuracies, therefore an error term $\delta$ is added.

The system (6.4, 6.5) can be integrated forward in time, starting from a state $\Psi_0$. Using Eq. (6.4) with $\epsilon = 0$, a preliminary state $\Psi_1^*$ is determined, and a forecast

$\Omega_1^* = \mathcal{G}(\Psi_1^*)$ (with $\delta = 0$) of the observables $\Omega_1$ at time $t = 1$ (Eq. (6.6)). This forecast deviates somewhat from the real observables, Thus, the eventual state $\Psi_1$ is estimated to be a linear combination of the preliminary forecast $\Psi_1^*$ and a term proportional to the error in forecasting the observable (Eq. (6.7)), i.e, $\mathcal{K}[\Omega *_1 - \Omega_1]$ with some suitable operator $\mathcal{K}$ (see, e.g., Robinson et al. 1998). This process is then continued in time.

$$\Psi_{t+1}^* = \mathcal{F}(\Psi_t, \eta) \tag{6.6}$$

$$\Psi_{t+1} = \Psi_{t+1}^* + \mathcal{K}\left[\Omega_t - \mathcal{G}(\Psi_{t+1}^*)\right] \tag{6.7}$$

This technique, which blends dynamical knowledge (the model $\mathcal{F}$) with empirical knowledge ($\Omega$) was first used in meteorological forecasting for improving the initial state, and is nowadays also a routine tool in ocean forecasting, and the operator $\mathcal{K}$ takes on a variety of advanced forms mostly derived from Kalman filters. It embeds mostly local observations.

In the case of improving regional detail in limited area meteorological modelling, the available empirical knowledge is given by the large-scale state of the atmosphere—the yellow part in Fig. 6.7. Thus, the (6.7)-step is done by correcting the large scale states—in meteorological applications given by Fourier components of the longest "waves".[9] The operator $\mathcal{K}$ is a weighting representing the trust in the large-scale components, which will be larger for larger scales and larger for vertical levels less influenced by surface properties such as land-sea contrast, mountains, etc.

The seminal paper that introduced and applied the concept of dynamical downscaling in climate simulations is credited to von Storch et al. (2000). This work involves a high-resolution reconstruction of Northwest European weather. The methodology was initially proposed by Waldron et al. (1996), though it did not initially receive the recognition it merited. Following the publication of the von Storch et al. (2000) paper, more research groups explored and employed the method, including studies by Miguez-Macho et al. (2004), Miguez-Macho et al. (2005), Radu et al. (2008), Omrani et al. (2012), and Alexandru et al. (2009).

In the following, we use, again, the case of simulating the strength and track of the East Asian typhoon Winnie Feser and von Storch (2008) for illustration.

From the state-space-approach (6.6,6.7) we expect several advantages ("added value"; Feser et al. 2011):

- reduction of internal variability;
- realistic geographical position of the large-scale features and their associated small-scale features (such as polar lows);
- meso-scale-improvements (such as intensity) of large-scale features.

[9] Either sine- and cosine terms or only cosine terms when using a cosine expansion for the limited area.

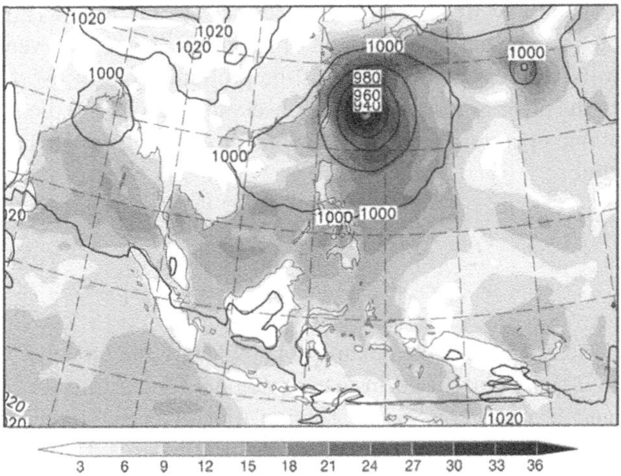


**Fig. 6.20** Simulation with spectral nudging of Winnie at 0000 UTC 17 Aug 1997 on the 50-km grid. The figure shows sea level pressure isobars (hPa) and shaded near-surface wind speed fields ($ms^{-1}$). © American Meteorological Society. Used with permission (Feser and von Storch 2008)

While we do not illustrate here the effect on the internal variability, we show in Fig. 6.20 the location of the typhoon "Winnie", which should be compared with the driving re-analysis shown in Fig. 6.18: There is only one storm, and it is geographically sitting correctly over Japan. Thus, the analysis follows the "empirical knowledge" provided by the driving re-analysis.

When comparing the intensity of the storm at different spatial scales (after digital filtering, Fig. 6.3). The two top maps in Fig. 6.21 show that the well-resolved part of the re-analysis is almost unaffected by the regional model. The distribution of the medium scales in the lower row of Fig. 6.21, however, are considerably smoother in the re-analysis (left), but steeper in the spectrally nudged regional model (right).

A distinct advantage of the downscaling approach is that the result is homogeneous in its skill across time, as long as the driving data are homogeneous in time when processing large-scale "predictors". This is probable because large-scale analyses based on coarse observational networks are relatively insensitive to improvements in the observational network.

Other cases demonstrating the achieved additional skill are describing the potential of the dynamical downscaling approach refer to North Atlantic Polar Lows (Zahn and von Storch 2008, 2010) and North Pacific Polar Lows (Chen and von Storch 2013; Chen et al. 2014), to "Medicanes" (Cavicchia et al. 2014), and low-level jets over the Yellow Sea area (Li et al. 2018). All these phenomena were at the time of the analysis too small to be resolved by the contemporary global re-analysis.

Further on, extended homogeneous reconstructions of the weather stream, including wind statistics, were done by Feser et al. (2001) for Northern Europe and for the Bo Hai and Yellow Sea (Li et al. 2016). These downscaled "re-analysis"

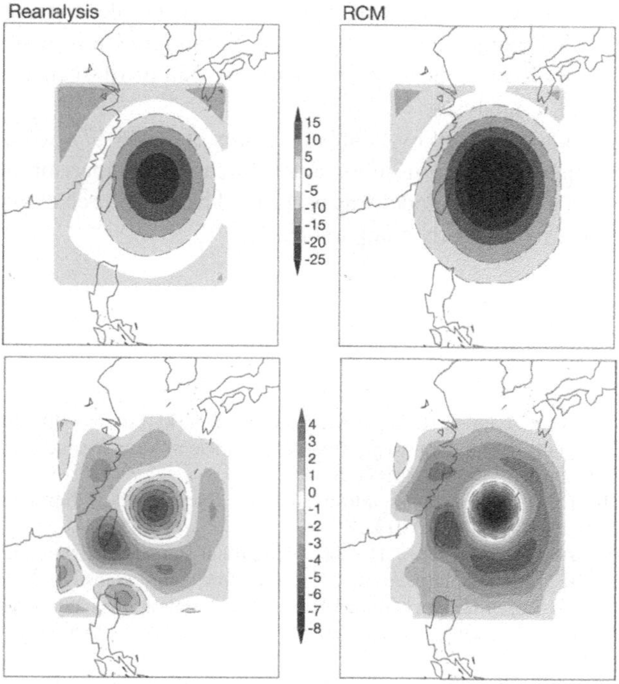


**Fig. 6.21** Spatially filtered air pressure deviations (hPa) at 0000 UTC 17 Aug 1997. (left) NCEP–NCAR reanalysis after interpolation on a 50-km grid; (right) RCM simulation with spectral nudging on a 50-km grid. (top) Large scales retained (diameter $\geq$ 600 km); (bottom) medium scales retained (about 180–360 km). Negative values are given as dashed isolines.). © American Meteorological Society. Used with permission (Feser and von Storch 2008)

were used, for instance, to force regional ocean models to derive statistics of storm surges (Woth 2005) and other marine impacts in the North Sea (Weisse et al. 2009).

Experiments with dynamical downscaling, employing spectral nudging in modelling regional oceans, have been done by Katavouta and Thompson (2013) and Katavouta and Thompson (2016) with encouraging results.

### 6.4.3 Global Modelling

An obvious, but long overseen option, of the dynamical downscaling has been suggested and demonstrated to be useful by Yoshimura and Kanamitsu (2008), who incorporated the spectral nudging element into a *global* atmospheric model. When dealing with climate change simulations with coupled atmosphere-ocean models, then a high spatial resolution of the atmospheric model must be associated with a similar grid resolution of the ocean model, and the spinning up of the coupled system may become prohibitively expensive. The same is valid for re-analysis, with a regular data assimilation scheme. Thus, the Yoshimura/Kanamitsu-Ansatz may be used as a cost-effective post-processor of the output of such coupled model simulations and of re-analyses.

The ansatz has been tested by Schubert-Frisius et al. (2017), who found added value mostly along coasts and in mountainous areas. A comparison of this data to downscaling with a regional model was performed by von Storch et al. (2017)—the results were found to be mostly comparable.

## References

Alexandru A, de Elia R, Laprise R, Separovic L, Biner S (2009) Sensitivity study of regional climate model simulations to large-scale nudging parameters. Mon Weather Rev 137(5):1666–1686. https://doi.org/10.1175/2008MWR2620.1

Benestad R (2016) Downscaling climate information. Res Encyclopedia Clim Sci. https://doi.org/10.1093/acrefore/9780190228620.013.27

Benestad RE, Hanssen-Bauer I, Chen D (2008) Empirical-statistical downscaling. World Sci. https://doi.org/10.1142/6908. https://www.worldscientific.com/doi/abs/10.1142/6908

Biau G, Zorita E, von Storch H, Wackernagel H (1999) Estimation of precipitation by kriging in the EOF space of the sea level pressure field. J Clim 12(4):1070–1085. http://www.jstor.org/stable/26244337

Branstator G (1995) Organization of storm track anomalies by recurring low-frequency circulation anomalies. J Atmos Sci 52(2):207–226. https://doi.org/10.1175/1520-0469(1995)052<0207:OOSTAB>2.0.CO;2

Busuioc A, von Storch H (2003) Conditional stochastic model for generating daily precipitation time series. Clim Res 24(2):181–195

Busuioc A, von Storch H, Schnur R (1999) Verification of GCM-generated regional seasonal precipitation for current climate and of statistical downscaling estimates under changing climate conditions. J Clim 12(1):258–272. https://doi.org/10.1175/1520-0442(1999)012<0258:VOGGRS>2.0.CO;2

Cavicchia L, von Storch H, Gualdi S (2014) Mediterranean tropical-like cyclones in present and future climate. J Clim 27(19):7493–7501. https://doi.org/10.1175/JCLI-D-14-00339.1

Chen F, von Storch H (2013) Trends and variability of North Pacific Polar Lows. Adv Meteorol 2013:170387. https://doi.org/10.1155/2013/170387

Chen F, von Storch H, Zeng L, Du Y (2014) Polar low genesis over the North Pacific under different global warming scenarios. Clim Dyn 43(12):3449–3456. https://doi.org/10.1007/s00382-014-2117-5

Cubasch U, Hasselmann K, Höck H, Maier-Reimer E, Mikolajewicz U, Santer BD, Sausen R (1992) Time-dependent greenhouse warming computations with a coupled ocean-atmosphere model. Clim Dyn 8:55–69

Davies HC (1976) A lateral boundary formulation for multi-level prediction models. Quarterly Journal of the Royal Meteorological Society 102(432):405–418. https://doi.org/10.1002/qj.49710243210

Denis B, Côté J, Laprise R (2002a) Spectral decomposition of two-dimensional atmospheric fields on limited-area domains using the discrete cosine transform (DCT). Mon Weather Rev 130(7):1812–1829. https://doi.org/10.1175/1520-0493(2002)130<1812:SDOTDA>2.0.CO;2

Denis B, Laprise R, Caya D, Côté J (2002b) Downscaling ability of one-way nested regional climate models: the big-brother experiment. Clim Dyn 18(8):627–646. https://doi.org/10.1007/s00382-001-0201-0

Feser F, von Storch H (2005) A spatial two-dimensional discrete filter for limited-area-model evaluation purposes. Mon Weather Rev 133(6):1774–1786. https://doi.org/10.1175/MWR2939.1

Feser F, von Storch H (2008) A dynamical downscaling case study for typhoons in Southeast Asia using a regional climate model. Mon Weather Rev 136(5):1806–1815. https://doi.org/10.1175/2007MWR2207.1

Feser F, Weisse R, von Storch H (2001) Multi-decadal atmospheric for Europe yields multi-purpose data. EOS Trans Am Geophys Union 82(28):305–310. https://doi.org/10.1029/01EO00176

Feser F, Rockel B, von Storch H, Winterfeldt J, Zahn M (2011) Regional climate models add value to global model data: A review and selected examples. Bull Am Meteorol Soc 92(9):1181–1192. https://doi.org/10.1175/2011BAMS3061.1

Frey-Buness F, Heimann D, Sausen R (1995) A statistical-dynamical downscaling procedure for global climate simulations. Theoret Appl Climatol 50(3–4):117–131. https://doi.org/10.1007/BF00866111

Fuentes U, Heimann D (2000) An improved statistical-dynamical downscaling scheme and its application to the alpine precipitation climatology. Theoret Appl Climatol 65(3):119–135. https://doi.org/10.1007/s007040070038

Giorgi F (1990) Simulation of regional climate using a limited area model nested in a general circulation model. J Clim 3(9):941–963. https://doi.org/10.1175/1520-0442(1990)003<0941:SORCUA>2.0.CO;2

Giorgi F, Mearns LO (1991) Approaches to the simulation of regional climate change: A review. Rev Geophys 29(2):191–216. https://doi.org/10.1029/90RG02636

Grotch SL, MacCracken MC (1991) The use of general circulation models to predict regional climatic change. J Clim 4(3):286–303. https://doi.org/10.1175/1520-0442(1991)004<0286:TUOGCM>2.0.CO;2

Karl TR, Wang WC, Schlesinger ME, Knight RW, Portman D (1990) A method of relating general circulation model simulated climate to the observed local climate. Part I: Seasonal statistics. J Clim 3(10):1053–1079. https://doi.org/10.1175/1520-0442(1990)003<1053:AMORGC>2.0.CO;2

Katavouta A, Thompson K (2013) Downscaling ocean conditions: Experiments with a quasi-geostrophic model. Ocean Model 72:231–241. https://doi.org/10.1016/j.ocemod.2013.10.001

Katavouta A, Thompson KR (2016) Downscaling ocean conditions with application to the Gulf of Maine, Scotian Shelf and adjacent deep ocean. Ocean Model 104:54–72. https://doi.org/10.1016/j.ocemod.2016.05.007

Kida H, Koide T, Sasaki H, Chiba M (1991) A new approach for coupling a limited area model to a GCM for regional climate simulations. J Meteorol Soc Japan Ser II 69(6):723–728. https://doi.org/10.2151/jmsj1965.69.6-723

Kim JW, Chang JT, Baker NL, Wilks DS, Gates W (1984) The statistical problem of climate inversion: Determination of the relationship between local and large-scale climate. Mon Weather Rev 112(10):2069–2077. https://doi.org/10.1175/1520-0493(1984)112<2069:TSPOCI>2.0.CO;2

Li D, von Storch H, Geyer B (2016) High-resolution wind hindcast over the Bohai Sea and the Yellow Sea in East Asia: Evaluation and wind climatology analysis. J Geophys Res Atmos 121(1):111–129. https://doi.org/10.1002/2015JD024177

Li D, von Storch H, Yin B, Xu Z, Qi J, Wei W, Guo D (2018) Low-level jets over the Bohai Sea and Yellow Sea: Climatology, variability, and the relationship with regional atmospheric circulations. J Geophys Res Atmos 123(10):5240–5260. https://doi.org/10.1029/2017JD027949

Maak K, von Storch H (1997) Statistical downscaling of monthly mean air temperature to the beginning of flowering of Galanthus nivalis L. in Northern Germany. Int J Biometeorol 41(1):5–12. https://doi.org/10.1007/s004840050046

Maraun D, Widmann M (2018) Statistical downscaling and bias correction for climate research. Cambridge University Press. https://doi.org/10.1017/9781107588783

Matulla C, Zhang X, Wang X, Wang J, Zorita E, Wagner S, von Storch H (2008) Influence of similarity measures on the performance of the analog method for downscaling daily precipitation. Climate Dyn 30(2):133–144. https://doi.org/10.1007/s00382-007-0277-2

Miguez-Macho G, Stenchikov GL, Robock A (2004) Spectral nudging to eliminate the effects of domain position and geometry in regional climate model simulations. J Geophys Res Atmos 109(D13). https://doi.org/10.1029/2003JD004495

Miguez-Macho G, Stenchikov GL, Robock A (2005) Regional climate simulations over North America: Interaction of local processes with improved large-scale flow. J Climate 18(8):1227–1246. https://doi.org/10.1175/JCLI3369.1

Omrani H, Drobinski P, Dubos T (2012) Spectral nudging in regional climate modelling: how strongly should we nudge? Q J R Meteorol Soc 138(668):1808–1813. https://doi.org/10.1002/qj.1894

Pielke RA (1991) A recommended specific definition of "resolution". Bull Am Meteorol Soc 72(12):1914–1914. https://doi.org/10.1175/1520-0477-72.12.1914

Radu R, Déqué M, SSomot (2008) Spectral nudging in a spectral regional climate model. Tellus A Dyn Meteorol Oceanogr 60(8):898–910. https://doi.org/10.1111/j.1600-0870.2008.00341.x

Robinson A, Lermusiaux P, Sloan III NQ (1998) Data assimilation. In: Brink KH, Robinson AR (eds) The global coastal ocean. Processes and methods. The sea, vol 10. Wiley, New York, pp 541–593

Roeckner E, von Storch H (1980) On the efficiency of horizontal diffusion and numerical filtering in an Arakawa-type model. Atmos Ocean 18(3):239–253. https://doi.org/10.1080/07055900.1980.9649090

Schubert-Frisius M, Feser F, von Storch H, Rast S (2017) Optimal spectral nudging for global dynamic downscaling. Mon Weather Rev 145(3):909–927. https://doi.org/10.1175/MWR-D-16-0036.1

Shumann FG (1957) Numerical methods in weather prediction: II. Smoothing and filtering. Mon Weather Rev 85(11):357–361. https://doi.org/10.1175/1520-0493(1957)085<0357:NMIWPI>2.0.CO;2

Starr V (1942) Basic principles of weather forecasting. Harper Brothers Publishers

Takayabu I, Kanamaru H, Dairaku K, Benestad R, von Storch H, Hesselbjerg Christensen J (2016) Reconsidering the quality and utility of downscaling. J Meteorol Soc Japan Ser II 94A:31–45. https://doi.org/10.2151/jmsj.2015-042

Tang S, von Storch H, Chen X (2020) Atmospherically forced regional ocean simulations of the South China Sea: Scale-dependency of the signal-to-noise ratio. J Phys Oceanogr 50:133–144. https://doi.org/10.1175/JPO-D-19-0144.1

Trenberth KE, Paolino DA (1980) The northern hemisphere sea-level pressure data set: Trends, errors and discontinuities. Mon Weather Rev 108(7):855–872. https://doi.org/10.1175/1520-0493(1980)108<0855:TNHSLP>2.0.CO;2

von Storch H (1995) Inconsistencies at the interface of climate impact studies and global climate research. Meteorol Z 4(2):72–80. https://doi.org/10.1127/metz/4/1992/72

von Storch H, Reichardt H (1997) A scenario of storm surge statistics for the German Bight at the expected time of doubled atmospheric carbon dioxide concentration. J Climate 10(10):2653–2662. https://doi.org/10.1175/1520-0442(1997)010<2653:ASOSSS>2.0.CO;2

von Storch H, Zorita E (2019) The history of ideas of downscaling – from synoptic dynamics and spatial interpolation. Front Environ Sci 7. https://doi.org/10.3389/fenvs.2019.00021

von Storch H, Zorita E, Cubasch U (1991) Downscaling of global climate change estimates to regional scales: An application to Iberian rainfall in wintertime. MPI report 64, Max-Planck-Institut für Meteorologie

von Storch H, Zorita E, Cubasch U (1993) Downscaling of global climate change estimates to regional scales: An application to Iberian rainfall in wintertime. J Climate 6(6):1161–1171. https://doi.org/10.1175/1520-0442(1993)006<1161:DOGCCE>2.0.CO;2

von Storch H, Langenberg H, Feser F (2000) A spectral nudging technique for dynamical downscaling purposes. Mon Weather Rev 128(10):3664–3673. https://doi.org/10.1175/1520-0493(2000)128<3664:ASNTFD>2.0.CO;2

von Storch H, Feser F, Geyer B, Klehmet K, Li D, Rockel B, Schubert-Frisius M, Tim N, Zorita E (2017) Regional reanalysis without local data: Exploiting the downscaling paradigm. J Geophys Res Atmos 122(16):8631–8649. https://doi.org/10.1002/2016JD026332

Waldron KM, Paegle J, Horel JD (1996) Sensitivity of a spectrally filtered and nudged limited-area model to outer model options. Mon Weather Rev 124(3):529–547. https://doi.org/10.1175/1520-0493(1996)124<0529:SOASFA>2.0.CO;2

Wang Y, Leung LR, McGregor JL, Lee DK, Wang WC, Ding Y, Kimura F (2004) Regional climate modeling: Progress, challenges, and prospects. J Meteorol Soc Japan Ser II 82(6):1599–1628. https://doi.org/10.2151/jmsj.82.1599

Weisse R, von Storch H, Mathcallies U, Chrastansky A, Feser F, Grabemann I, Günther H, Pluess A, Stoye T, Tellkamp J, Winterfeldt J, Woth K (2009) Regional meteorological–marine reanalyses and climate change projections: Results for Northern Europe and potential for coastal and offshore applications. Bull Am Meteorol Soc 90(6):849–860. https://doi.org/10.1175/2008BAMS2713.1

Whetton PH, Katzfey JJ, Hennessy KJ, Wu X, McGregor JL, Nguyen K (2001) Developing scenarios of climate change for southeastern Australia: an example using regional climate model output. Climate Res 16(3):181–201. https://doi.org/10.3354/cr016181

Wigley TML, Jones PD, Briffa KR, Smith G (1990) Obtaining sub-grid-scale information from coarse-resolution general circulation model output. J Geophys Res Atmos 95(D2):1943–1953. https://doi.org/10.1029/JD095iD02p01943

Woth K (2005) North sea storm surge statistics based on projections in a warmer climate: How important are the driving GCM and the chosen emission scenario? Geophys Res Lett 32(22). https://doi.org/10.1029/2005GL023762

Yoshimura K, Kanamitsu M (2008) Dynamical global downscaling of global reanalysis. Mon Weather Rev 136(8):2983–2998. https://doi.org/10.1175/2008MWR2281.1

Zahn M, von Storch H (2008) A long-term climatology of North Atlantic polar lows. Geophys Res Lett 35(22). https://doi.org/10.1029/2008GL035769

Zahn M, von Storch H (2010) Decreased frequency of North Atlantic polar lows associated with future climate warming. Nature 467(7313):309–312. https://doi.org/10.1038/nature09388

Zorita E, von Storch H (1999) The analog method as a simple statistical downscaling technique: Comparison with more complicated methods. J Climate 12(8):2474–2489. https://doi.org/10.1175/1520-0442(1999)012<2474:TAMAAS>2.0.CO;2

Zorita E, Kharin V, von Storch H (1992) The atmospheric circulation and sea surface temperature in the North Atlantic area in winter: Their interaction and relevance for Iberian precipitation. J Climate 5(10):1097–1108. https://doi.org/10.1175/1520-0442(1992)005<1097:TACASS>2.0.CO;2

Zorita E, Hughes JP, Lettemaier DP, von Storch H (1995) Stochastic characterization of regional circulation patterns for climate model diagnosis and estimation of local precipitation. J Climate 8(5):1023–1042. https://doi.org/10.1175/1520-0442(1995)008<1023:SCORCP>2.0.CO;2

# Chapter 7
# Paleoclimate and Historical Climate Reconstructions

**Eduardo Zorita, Hans von Storch, Martin Claussen, Martin Heimann, and Robert Sausen**

**Abstract** Climate research is constrained by the practical impossibility of experimenting with the climate system to test hypotheses and the limited length of observations, which usually hampers a thorough statistical analysis. This is one of the reasons why paleoclimatology, the study of past climate variations can give valuable insights about present climate change and put this change against a broader background. However, palaeoclimatology needs to rely on indirect evidence, so-called *proxy data*, which contains information about past environmental conditions and needs to be adequately interpreted. Often, this interpretation requires rather sophisticated statistical analysis. This chapter presents four important contributions to paleoclimatology that arose within the Hamburg climate network. Three of them analyze available indirect observations to infer past climate variations in the recent past and also at longer glacial timescales. The fourth contribution is more methodological and focused on ways to test statistical climate reconstruction methods using synthetic data.

One of the main limitations of climate research is the impossibility of carrying out experiments under controlled conditions, as other areas of natural science routinely do in a laboratory. Therefore, it heavily relies on analysing direct meteorological and marine observations. This statistical analysis is often hampered by small sample sizes, errors in observations, inhomogeneous data sets originating from different, or not perfectly controlled, measuring devices, etc. All those data imperfections impact the conclusions reached from their statistical analysis. As we have just one realisation of the climate trajectory and the sample size does not allow us to set aside a validation data set, that impact cannot be easily quantified. To expand the sources of climate information and to cover longer time scales that could be possible using just observations, climatology also considers other types of data that may contain indirect information about present and past climate conditions. Well-known examples of proxy data are tree-ring widths, as trees grow wider rings in years with more favourable conditions, e.g. warmer temperatures or higher precipitation. Analyzing those data can provide valuable information about past temperatures, precipitation, ocean currents, etc. Those indirect indicators (e.g. tree-ring widths)

 
H. von Storch et al., *Climate Science Concepts Born in Hamburg*,
https://doi.org/10.1007/978-3-031-81108-1_7

need, however, to be quantitatively, or at least qualitatively, interpreted in terms of climatological variables (e.g. temperature deviations from mean conditions), also possibly interpolated and perhaps extrapolated to reconstruct complete climate fields from the available sparse network of proxy data.

In this chapter, we review a few important contributions in the field of paleo-climate and paleoclimate reconstructions that arose within the Hamburg research network. Three of these contributions interpreted proxy data to provide valuable information about past climate variations in the North Atlantic and Baltic Sea—the third and fourth contributions aimed at providing an analytical framework to test climate reconstruction methods.

## 7.1 Heiner Schmidt's Geostrophic Proxies for Marine Storminess

Since the 1980s, there has been a growing interest in whether or not climate change has led to an increase in the frequency or intensity of storms, as indicated by a series of studies (e.g., Neu 1984; Carter and Draper 1988; Hogben 1994; von Storch et al. 1993). The most obvious way of assessing such change would be to rely on observations of storm damages or series of weather maps—but these almost always suffer from inhomogeneities, be it by changes in the location and environment of the measuring instrument, a change of the observer (in particular, when visual assessments according to the Beaufort-scale are archived) or when other factors influence the recorded data, such as damages, in an uncontrolled manner (Karl et al. 1993; Pielke and Landsea 1998; WASA group 1998; Lindenberg et al. 2012). Often, the problem can be dealt with by techniques known as "homogenization", but hardly so when dealing with wind observations, since such inhomogeneities not only consist of sudden changes or breaks but of irregular continuous disturbances that may extend over a specific time span as well.

Heiner Schmidt (Fig. 7.1) from the Seewetteramt in Hamburg, part of the German Weather Service, developed a sound solution to detect real changes in storminess by examining changes in geostrophic wind statistics (Schmidt and von Storch 1993). In a rotating system, the wind is subject to the Coriolis force perpendicular to its velocity. The geostrophic wind is defined as the wind velocity and orientation whose Coriolis force exactly opposes the air pressure gradient. This is denoted as geostrophic equilibrium. Wind averages over time spans of one day or longer and at spatial scales of hundreds of kilometres tend to be in geostrophic equilibrium with air pressure. This equilibrium dictates that winds tend to flow approximately parallel to the isobars in weather maps. Therefore, the geostrophic wind can be approximately inferred from air pressure readings. There are several advantages of focusing on pressure readings instead of wind observations: pressure time series usually cover a more extended period than wind observations; they are also less prone to measurement errors, considering the rather erratic and turbulent behaviour

**Fig. 7.1** Heiner Schmidt (left), who has introduced the usage of changing geostrophic wind statistics as proxy for changing storminess together with Oliver Krüger (right), who provided evidence for the power of this argument. Foto: Hans von Storch

of wind. Air pressure is almost unaffected by the environment of the measuring device, such as trees, buildings, etc., that do affect wind recordings. Schmidt aimed to identify long-term changes in wind statistics in the German Bight, particularly statistics of extreme winds and decided to use long pressure records for the reasons just mentioned. He identified three locations with continuous pressure readings and, if data were unavailable, used readings from nearby stations (see Fig. 7.2). The data were quality-controlled and checked for inhomogeneities, but air pressure is not particularly sensitive to changes in observational practices. From the triangle of pressure readings (Fig. 7.2), a geostrophic wind was determined, here for the

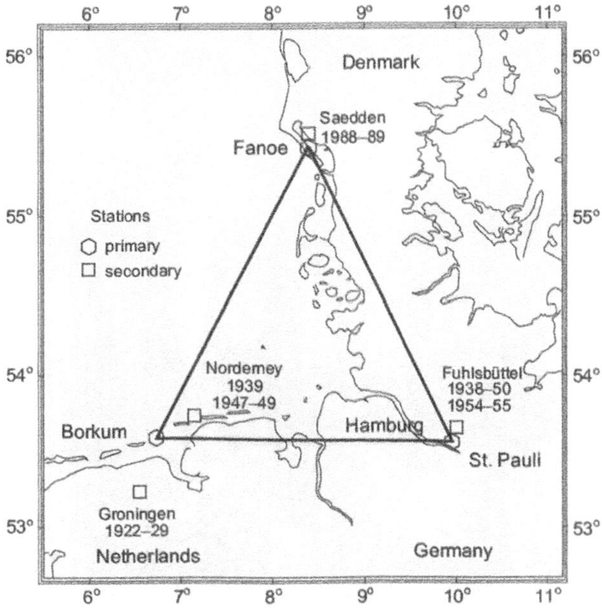


**Fig. 7.2** Heiner Schmidt used daily pressure readings from the three corners of a triangle and calculated the geostrophic wind speed for the German Bight from these pressure readings. From Schmidt and von Storch (1993)

German Bight, for each day from 1876 until 1989. Then, for each year, the distribution of all daily values was ordered, and that value larger than *p* % of all values—the *p*th percentile—was determined for different values of *p*. Despite the increasing efficiency of the greenhouse effect towards the end of the series, none of the time series of annual percentiles exhibited an increasing trend, particularly not an accelerating one. Figure 7.3 shows one of those time series, namely the 99% percentile.

By utilising additional data and integrating various estimates, the methodology was subsequently broadened (Alexandersson et al. 1998; Matulla et al. 2008; Wang et al. 2011; Matulla et al. 2012; Krueger et al. 2019), resulting in a conclusive outcome that entirely aligned with earlier findings: "The results show that storm activity over the German Bight [as well as over Northern and Central Europe], is subject to multidecadal variability. The latest decades are characterised by increased activity from the 1960s to the 1990s, followed by a decline lasting into the 2000s and below-average activity up until the present." (Krieger et al. 2021).

The approach's effectiveness hinges on the correlation between fluctuations in wind speed statistics and those of geostrophic wind speeds. This assumption was validated by Oliver Krüger (Fig. 7.1), who demonstrated the consistency of variations in wind and in geostrophic wind statistics using data from a multi-decadal simulation with a climate model (Krueger and von Storch 2011). This is

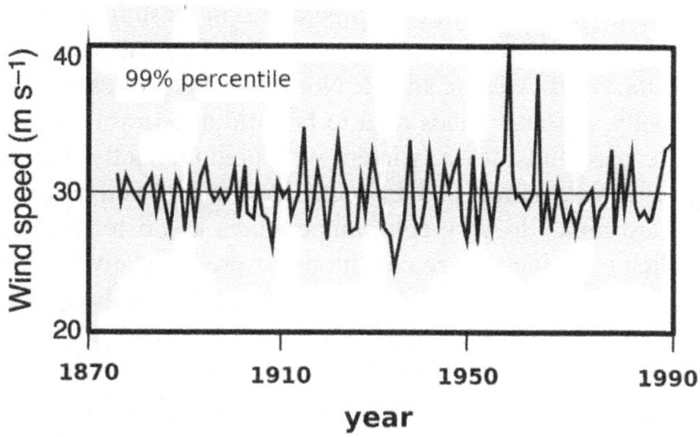


**Fig. 7.3** Annual time series of the 99% percentile of the daily geostrophic wind in the German Bight derived from triangulation of air-pressure readings using the geostrophic approximation (see Fig. 7.2). Modified from Schmidt and von Storch (1993)

the first example of using synthetic data from climate models to test and validate reconstruction methods. A more extensive application of this idea is the focus of Sect. 7.4.1 later in this chapter.

Moreover, the method was utilized to gauge the dependability of an extension of the atmospheric state ("re-analysis") during the first half of the twentieth century, a period with no available upper air data (Krueger et al. 2013). To accomplish this, Oliver Krüger computed wind statistics and their fluctuations in the re-analysis for sites where well-documented observed air pressure data were available. However, the trends generated by the reconstructions were found to be inconsistent with empirical data based on the geostrophic wind speed statistics—a problem likely addressed in more recent centennial reanalysis datasets.

## 7.2 Baltic Sea Ice Cover

The reconstruction of Baltic Sea Ice cover over the past centuries represents an excellent example of the analysis of indirect observations—not properly meteorological observations, but not indirect proxy data. This reconstruction is based on "historical evidence" from chronicles, administrative and religious records, commercial contracts, private letters, port reports of sailing seasons, accounts of transport of goods over sea-ice, etc. These documents need to be carefully searched in historical archives, filtered according to their climatic relevance, their reliability assessed and then comprehensively analyzed and summarized to construct a numerical regional climate index of ice severity, in this case, of the frequency and intensity of sea-ice cover in the Western Baltic Sea. This variable is climatologically important, not only because it influences the Baltic ecosystem as such, but also because it is a sensitive indicator of the temperatures in Western Europe over the past centuries, and it can even be interpreted as an indirect indicator of the atmospheric circulation over the North Atlantic and the North Sea. In the present climate, winters with predominantly westerly winds tend to be mild and stormy for the North and Western Baltic coasts. In contrast, winters with predominantly meridional (north-south) atmospheric circulation and frequent *blocking* situations over the central North Atlantic are linked to very cold temperatures and extensive ice cover. The first reconstruction of Baltic sea-ice conditions covered a relatively short time span, since 1701 (Koslowski and Loewe 1994). These authors subsequently expanded it to cover the period from 1500 onwards (Koslowski and Glaser 1999). Then, Schmelzer and Holfort from the "Bundesamt für Seeschifffahrt und Hydrographie" in Hamburg published an expansion of the period covered by the index since 1301 (Schmelzer and Holfort 2011), a remarkable achievement, based on the two previous reconstructions, considering the complexity of compiling and interpreting the historical sources from those early centuries. The reconstructed Western Baltic Sea Ice cover index is displayed in Fig. 7.4. The index indicates less extensive ice cover in the earlier and more recent centuries, with more extensive sea-ice cover during the Little Ice Age, between about 1500 and 1800. This is consistent with temperature reconstructions for Europe based on dendroclimatological data. The reasons for those fluctuations in temperature and sea-ice cover may lie in the variations of the external climate forcing, particularly the frequency and intensity of volcanic eruptions and the sun's output. Both contributed to lowering the mean temperatures during the Little Ice Age (PAGES 2k Consortium 2013).

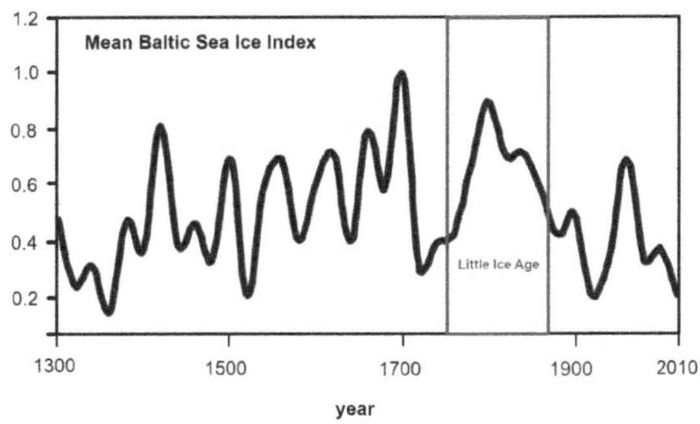


**Fig. 7.4** Reconstructed sea-ice severity index for the Western Baltic Sea (figured derived from Schmelzer and Holfort (2011) and modified) based on historical evidence. The units are semiquantitative: very weak and weak (index = 0), moderate– (0.1), moderate+ (0.3), strong– (0.5), strong+ (1), very strong (2), and extremely strong (3) ice winters. As the series is smoothed with a 40-year low-pass filter, extremely strong ice severity in individual winters is not resolved in this figure. The rectangle with blue borders identifies the duration of the Little Ice Age, according to Schmelzer and Holfort

## 7.3 Hartmut Heinrich's Stone and the Heinrich Events

In paleoclimatology, the term *Heinrich events* is now widely recognised, referring to brief spells—at geological timescales- of cold climatic conditions that quasi-periodically occurred in the past seven glacial phases. The Earth's climate history of the past million years is characterised by quasi-oscillations between ice ages (glacial stage) that approximately last 100 thousand years, separated by short warmer phases, like the current stage, that last about 10 thousand years. During the glacial phases of these cycles, large parts of the Northern Hemisphere land areas at high latitudes are covered by ice sheets that may attain a thickness of a few kilometres only to recede and eventually almost wan, except for Greenland, during the shorter *interglacial phases*.

We know now that during the glacial phases of the Pleistocene (Hodell et al. 2022), a massive number of icebergs, literally an *armada* of them, were intermittently released mostly, but not only, from the Laurentide Ice Sheet in Northern North America. These are the *Heinrich events* (Bond et al. 1992). This armada then sailed through the Hudson Strait into the North Atlantic and eventually melted. The significance of these Heinrich events lies in the large amount of freshwater carried to the North Atlantic by the melting icebergs. This sudden influx of cold and fresh water has the potential to alter one of the drivers of circulation of the North Atlantic, determined by the gradients of temperature and salinity, the so-called *thermohaline circulation*, which could cause significant global climate variability. Future changes in this circulation caused by the warming of the Atlantic surface layers could also be relevant in the context of climate change.

The glacial stages are punctuated by the so-called *Heinrich events*. The discovery of the Heinrich events is illustrative for a wider readership not only because of its huge scientific importance in palaeoclimatology, but also because of the *human story* attached to its discovery. It shows that science is not always carried out in an ivory tower isolated from worldly matters but that it is often driven by individual curiosity, sometimes also by chance, and wrapped in more mundane human passions. This is our account of it.

Although the term "Heinrich events" is widely used in paleoclimatology, very few people are currently aware of who this person, Hartmut Heinrich, is. Despite working at the "Bundesamt für Seeschifffahrt and Hydrographie" (BSH) in Hamburg, he remained relatively unknown in Germany. However, this changed when a detailed account of Heinrich's discovery, in his own words, was published in an interview (von Storch and Emeis 2017).

It all started when Hartmut Heinrich discovered a stone hoisted from the bottom of the sea bed that one of those icebergs had transported to the deep sea off the Iberian coast. But he did not only discover the stone, and found it strange, he also interpreted it: the icebergs that caused these events carried rock material that must have been eroded by glaciers, which must have been deposited onto the sea floor as ice-rafted debris. In his own words: "In the mid-eighties, the German Hydrographic Institute (DHI) carried out an environmental impact assessment on the deposition

**Fig. 7.5** Hartmut Heinrich in 2016 with his stone in his office at Bundesamt für Seeschiffahrt und Hydrographie in Hamburg. Foto: Hans von Storch

of intermediate to high-level radioactive waste .... The assessment region lay in the eastern foothills of the Mid-Atlantic Ridge ... with water depths ranging from 3700 to 4500 meters." A ship was sent out, with the young geologist Hartmut Heinrich on board, who the DHI had hired following the advice of the employment agency.[1]

Heinrich **read** the "the stone" (Fig. 7.5) and learned: "... box corer samples were taken from the upper fifty centimetres of the seabed. A chunk of basalt measuring twenty centimetres was randomly found in one of the box corers. It was a glacigenically[2] faceted Icelandic basalt. The upper and lower parts were covered with black manganese oxide, and in the middle was a bright strip two to three centimetres in width; above and below was a band of reddish-brown iron oxide. This typical characteristic is observed on stones partially embedded in oxygen-free sediments and partially protruding into oxygen-containing bottom water. The bright stripe, therefore, should represent a layer of oxygen-free sediment. The sediment layer in this corer, which bordered against this bright strip, differed from the under and overlying brown sediment due to its very bright colour". [However, the anoxic

[1] Later, the DHI would rename itself to BSH.

[2] Polished and moulded by the action of flowing ice.

condition did not originate from the organic carbon content, which was about half of that in the under- and overlying oxic sediments.]:

"... My conclusion was that northeastern Atlantic bottom water must have been anoxic during the deposition of this layer..... the sand fraction in this layer consisted nearly exclusively of crystal-clear, sharp-edged quartz grains and the few foraminifera[3] found belonged to a species that today only exists north of the polar front. The layer was somewhat diagenetically solidified by a mineral that often forms in anoxic situations, namely dolomite. Obviously, this stone sunk from a thawing iceberg, which likely had drifted from Iceland to the site where we discovered it. Sometime later, the *ominous* layer's deposition took place over a relatively short period, followed again by the *normal* ice age conditions in the northeastern Atlantic as before."

The term *Heinrich events* was actually coined by Jerry McManus, affiliated with the Lammont Doherty Earth Observatory. Another well-known scientist, Wally Broecker, soon began using the term as well (Broecker et al. 1992). At the 1992 Kiel Paleoceanography Conference, over a dozen papers referred to the "Heinrich events", all authored by American, British, and French scientists. However, German academia seemed to wrestle with accepting this terminology: "..One of the [German] top dogs alleged that he had made the discovery himself, which no one believed. I recently found out that he had been trying for fifteen years to prove that it had no substance. Otherwise, the contribution has largely been ignored in Germany. Abroad, it spread like wildfire, though. There was a great deal of international approval. The events are a worldwide component of geoscience curricula today, and many students are surprised when they find out I'm still alive". Also, the scientific establishment in Hamburg showed no interest in this fellow at the Bundesamt für Seeschifffahrt und Hydrographie, allegedly because he had not graduated in Hamburg.

## 7.4 Testing Climate Reconstruction Methods

In the previous sections, we have seen a few contributions of reconstructions of some characteristics of past climates based on indirect information recorded in natural archives that can be read today. This framework is also applied to many other types of proxy data, comprising tree-rings, lake sediments, corals, speleothems, and many others. These natural archives only record past environmental conditions indirectly and are also affected by other non-climatic factors that blur the climate signal we wish to extract from them. In practice, more sophisticated statistical methods combine all these data types to derive a complete map—not only one particular characteristic- of past temperatures or precipitation for past periods. For example, with their high and accurate annual resolution, tree-ring data would ideally allow

[3] Single cell, skeleton-building organisms that mostly dwell on the sea-bed.

the reconstruction of a map of summer mean temperatures over the whole globe. Obviously, these reconstructions are burdened by uncertainties stemming from the limited coverage of the proxy network and the *noisy* characteristics of the data derived from natural archives. A method to quantify and test the reconstruction methods is required. One may think there is no easy solution to the problem of limited data and poor data quality. If the information has not been collected or is of poor quality, it does not exist and cannot be just generated out of a hat. However, there is a way to quantify those problems' impact on the statistical analysis's robustness and estimate the uncertainties inherent in incomplete or imperfect data analysis. It is possible to figure out ways to optimise the data collection processes to diminish those statistical uncertainties.

These methods are based on *synthetic* data: data that can be relatively easily produced, without temporal or spatial gaps, with no measurement errors, and which statistically and physically resemble actual observations. These data would represent an ideal synthetic laboratory where the statistical analysis could be recreated, testing the underlying assumptions in *numerical experiments*, and comparing the conclusions derived from the analysis with this *virtual reality*. Those 'magical' data can be generated with suitable transformations from climate simulations with Earth System models.

Although Earth System models do not perfectly represent reality, they do encapsulate our knowledge of the dynamics of the climate systems. The simulations conducted with those models produce climate projections, which are already well-known. To a reasonable extent, they are physically consistent, comprising many physical variables without spatial or temporal gaps. Apart from projecting future climate change, Earth System Models can also simulate the present and past climate over much longer than the twentieth century. They represent a *digital twin* of the Earth, albeit with a discreet spatial and temporal resolution. Climate data from climate simulations can be considered a synthetic laboratory to test a given statistical analysis: if statistical analysis fails in those laboratories, they will likely also fail in the more complex system of the real Earth.

This idea of synthetic data first crystallised to test methods to reconstruct past climates (von Storch et al. 2004; Mann and Rutherford 2002). These methods are based on proxy data (such as the well-known tree-ring widths). Tree-ring width and other biological indicators can represent environmental conditions (temperature, precipitation) in the past with annual temporal resolution over periods of several millennia. This representation is, however, imperfect, as many other non-climatic factors can also influence tree growth. These confounding factors that are not necessarily related to climate variations comprise, for instance, wildfires, competition with other trees, parasites, and nutrient availability, and all represent a source of nuisance variations, denoted by the term *noise*. The measured tree-ring widths also need, first, to be translated to temperature or precipitation units. In the second step, the collection of data from a network of proxy sites is used to reconstruct the whole temperature (or precipitation) field over the entire Northern Hemisphere (see Fig. 7.6). But before the statistical machinery of a statistical climate reconstruction method can be tested, suitable synthetic data must be produced.

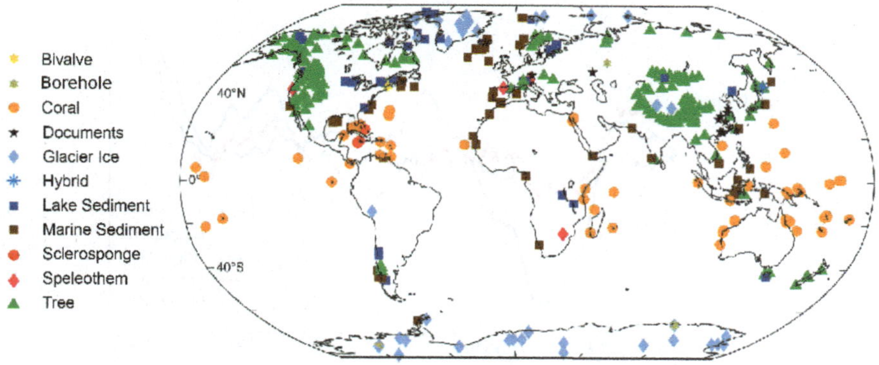


**Fig. 7.6** An example of a network of sites with proxy records, with the proxy type, used to reconstruct past temperatures over the past two millennia (based on PAGES 2k Consortium 2013). Each proxy type (and at each site) may be sensitive to environmental conditions, e.g. mean growth-season temperature, annual precipitation, etc., over a particular season and be representative of a particular region. Therefore, the climate information in the proxy records is heterogeneous and contaminated by non-climatic sources of variations. Statistical methods are needed to translate this information into complete global climate fields

## *7.4.1 Construction of Pseudo-Proxies*

The global or large-scale past temperature reconstruction relies on individual series of proxy records that indirectly represent local temperature or precipitation. An example of these temperature proxy time series is presented in Fig. 7.7. The series represent smoothed versions of tree-ring-derived series from Central Europe, two historical records of the start of the grape harvest in France and Switzerland, and instrumental temperature series from a meteorological station close to the proxy sites and the temperature averaged over France. The figures show clear agreements between all these series, but also clear deviations as well. These deviations are caused by several non-climate factors that may affect the grape harvest dates or the tree-ring growth. When the tree ring and the date-of-harvest series are used, along with other proxy series, to reconstruct larger-scale temperatures, these deviations cause uncertainties and biases.

The testing of climate reconstruction methods that rely on surrogate pseudo-proxy records should, therefore, produce surrogate pseudoproxy series that mimic the characteristics of real proxy records. These characteristics must also include the deficient connection between proxies and the instrumental temperature series. There are two ways to produce realistic but synthetic proxy time series, both based on data from climate simulations.

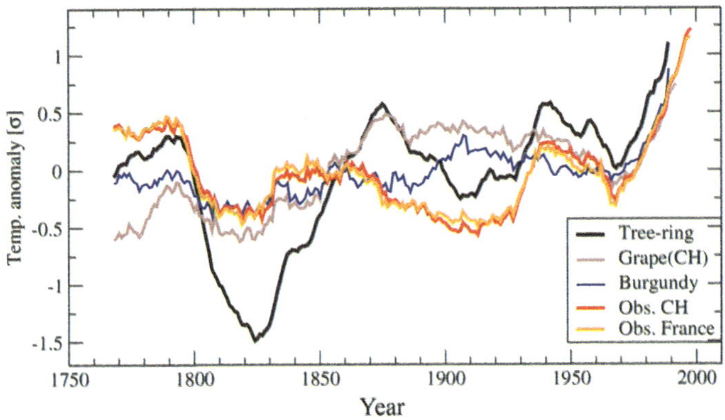


**Fig. 7.7** An example of real proxy series and their corresponding instrumental temperatures. The proxy records are derived from tree-ring measurements located in Central Europe and from historical accounts of the start of the grape harvest in Burgundy (France) and in Switzerland. The instrumental temperature series represents the measured mean temperature in Switzerland and France. The series have been aligned to represent deviations from their long-term mean, re-scaled to unit standard deviation, and smoothed with a low-pass 30-year filter. The typical correlations between an unsmoothed (annual) proxy series and the corresponding instrumental temperature series are usually in the range of 0.3–0.5, with the best proxies attaining correlations of 0.7. The figure is from Ludescher et al. (2020)

## *7.4.2 Statistical Methods*

One method, the most widespread, is based on building statistical models of those time series. These synthetic temperature or precipitation records can be constructed from climate simulations in two steps. The first is by taking the temperature or precipitation simulated by the Earth System model in each model grid cell. The second step is to contaminate those simulated temperature time series with the required amount of statistical noise $\epsilon$ so that the correlation between the synthetic temperature $T$ and the pseudo-proxy $pp$ attains a previously specified value $r$:

$$pp(t) = T(t) + \epsilon(t). \tag{7.1}$$

Here, the amplitude (standard deviation) of the noise term $\epsilon$ can be derived from the intended value of the correlation $r$.

However, the question is more complex than it may seem at first sight, as Fig. 7.7 already hints at the noise term $\epsilon$ representing the non-climatic noise present in the real proxies. This noise can have rather peculiar statistical characteristics, particularly related to the temporal memory or *autocorrelation*. For instance, the growth of trees may also depend on nutrients stored in the previous growth season; the impact of parasites or fires may slowly fade over several years; the physiological damage caused by drought may mark the tree growth over a long time. This autocorrelation can impact a statistical model that links the same-year values of proxy and temperature. Therefore, the pseudo-proxies should also incorporate these characteristics of the statistical memory to test the reconstruction method in a realistic setting. However, these statistical characteristics cannot be easily represented and usually are rather specific to the proxy location and type. The usual approach when testing reconstruction methods is to construct pseudo-proxies with a wide range of possible statistical properties so that a worst-case scenario can be accounted for.

Beyond the amplest assumption that the noise in the pseudo-proxy can be represented by a random, normally distributed, and serially uncorrelated time series (white Gaussian noise), the noise term can also be represented by an autocorrelated (*red*) processes and by *long-term-memory* processes (also denoted as a FARIMA process, Autoregressive fractionally integrated moving average). In a red noise series, a time step is explicitly (independently) correlated with a finite number of previous time steps. Generating a red-noise process is technically relatively simple by applying an autoregressive process of low order. In a long-term-memory process, a time step is (independently) correlated to all other previous—in theory infinite—time steps (Ludescher et al. 2020). The result is that the serial correlation diminishes more slowly with time lag than in a red-noise process. Generating a long-term memory process is technically considerably more complex than an autoregressive process (Makse et al. 1996; Liu et al. 1999). Therefore, these two types of processes represent two limiting cases that can be used to generate noise in the synthetic pseudo proxies. The noise in real proxies probably lies between these two limiting cases regarding serial autocorrelation. By generating surrogate proxies in these two limit cases, one can estimate the range of uncertainties and biases resulting from serial correlation in the real proxies. In general, the presence of a long-lag serial correlation in the non-climate noise causes a deterioration of the reconstructions, with larger biases and a more pronounced underestimation of past climate variability (von Storch et al. 2009).

### 7.4.3 Forward Modelling of Proxies

An alternative, more sophisticated method to generate pseudo-proxies is applying a proxy *forward model*. These models incorporate the biological and geochemical knowledge of the most relevant processes influencing the proxy indicator. For instance, a tree-ring forward model would simulate the growth of a tree-ring in a particular growing season using input values of temperature, solar irradiance, available humidity, etc. (Tolwinski-Ward et al. 2011). Figure 7.8 displays a schematic representation of the tree-ring forward model.

A forward model of the isotopic composition of an ice-core would take as input the isotopic composition of water vapour and simulate the fractionation that occurs in the phase transitions from gas to liquid to solid water (Sturm et al. 2010; Klein et al. 2019), and similarly for as stalagmite forward model (Baker et al. 2012). The construction of pseudo proxies has even been applied to ecological systems (Asena

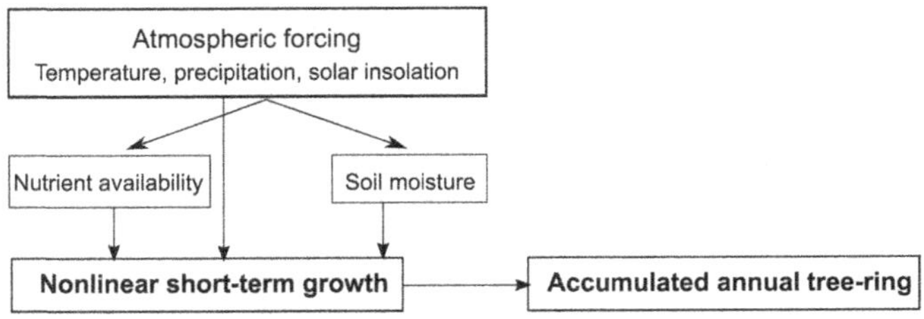


**Fig. 7.8** A schematic representation of a forward model of tree-ring growth. The model takes as input the meteorological forcing and translates this forcing to the growth of a tree ring through a set of simplified equations of tree biochemistry. This model does not, in principle, include some source of non-climatic noise, such as parasites or wildfires and also assumes that the meteorological forcing is measured at the proxy site. However, it represents more faithfully than a statistical model of possible non-linear relationships between forcing and tree growth. This figure is a strongly simplified scheme of the VS-Lite model (Tolwinski-Ward et al. 2011)

2021). Forward modelling. The level of sophistication of these models can vary. Some of them would use monthly mean values of the environmental drivers of tree-ring growth, and others would resolve their daily cycle, as many of the processes involved in the generation of a proxy record are non-linear and, therefore, require a high temporal resolution for a better representation of their variability.

### 7.4.4 *Paleoclimate Pseudo-Reconstruction*

The statistical steps needed to complete a paleoclimate reconstruction may comprise a rather complex chain of statistical methodologies to extract the climate signal from the non-climatic noise and extrapolate a sparse network of point-wise data to a whole large-scale field. In the end, it is difficult to robustly estimate the biases and uncertainties of these methodologies based solely on a limited period in which actual observations exist.

An important application of pseudo-proxy tests (von Storch et al. 2004) arose in the context of temperature reconstructions over the past millennium, specifically the "hockey-stick" reconstruction (Mann et al. 1998; National Research Council et al. 2007). This type of climate reconstruction usually investigates whether the temperature in the past surpassed the current temperature level, which is strongly driven by the concentrations of atmospheric greenhouse gases. Another objective of the reconstructions is to estimate the amplitude of multidecadal temperature variations. The hockey-stick reconstruction was based on a collection of Northern Hemisphere proxies. It resulted in a very flat temperature evaluation over the pre-industrial centuries and a sharp and sudden warming thereafter. The pseudo-proxy tests, however, indicated that the reconstruction method artificially flattens the amplitude of past temperature variability. This result displayed the need to design alternative climate reconstruction methods, but also, more importantly, highlighted the requirement of rigorous testing of climate reconstructions methods, and other statistical methods applied in climatology, under controlled conditions (Christiansen et al. 2009; Smerdon 2012).

Although climate reconstructions over the past millennium that were produced in the 2000s showed common features, mainly the dominant warming trend at a global scale in the twentieth century, they also presented dissimilarities in the amplitude of past climate variations and the evolution of temperature in certain regions, for instance, the Tropical Pacific. The reason for this divergence would be initially assumed to lie in the different sets of proxy data used to reconstruct the climate. Still, several studies pointed to another source, namely the *fingerprint* in the reconstruction methods. As mentioned before, the statistical machinery involved in the reconstructions can quickly become rather convoluted, ranging from the relatively simple Principal Components Regression through Regularized Expectation Maximization (Rutherford et al. 2005) and Bayesian Hierarchical Modelling (Tingley and Huybers 2010), to name a few prominent examples. Even within a well-established method, researchers sometimes need to make subjective decisions on particular details of the technique, such as truncation of the dimensionality. It turned out that these decisions had a relativity strong impact on the final reconstructions and that different equally justifiable decisions would produce distinct reconstructions (Bürger and Cubasch 2005; Christiansen et al. 2009). In addition, the reconstruction skill of a particular method was found to also be strongly dependent on the specific data set used to evaluate it. For instance, additional data sets generated from the same climate model, which in principle would provide a similar assessment of the skill of a particular method, yet yielded different evaluations (Christiansen et al. 2009). They were too strongly dependent on the specific data sample. This indicated that a reconstruction method needed to be assessed in an ensemble of samples, not just in one, even generated by the same climate mode. This somewhat surprising result would have remained unnoticed if only the observational record had been used to evaluate one reconstruction method.

An example of the main effects of noise in the proxy series on climate reconstructions is displayed in Fig. 7.9. The figure shows the temporal correlation between the simulated and pseudo-reconstructed temperatures from a global climate simulation over the past millennium. The locations of the network of pseudo-proxies mimic a real network of proxies, and the amount of noise included in the pseudo-proxies is either zero (Fig. 7.9a) or realistic (derived from the comparison of real proxies series with collocated real temperature measurements). The correlations are high but never reach the value of one, even with perfect (no-noise) pseudo-proxies. This indicates that the objective of reconstructing the global temperature field from the spatially limited information in a set of sites cannot be ultimately attained. It is even more complicated when the pseudo-proxy is constructed to contain a realistic amplitude of non-climatic noise: the correlation to the model-simulated diminishes, and high values of these correlations are restricted to areas with a dense collection of proxy sites (see Fig. 7.6). Although this result is generally expected, pseudo-proxies experiments help quantify the loss of skill that can be attributed or expected from noisy proxy records.

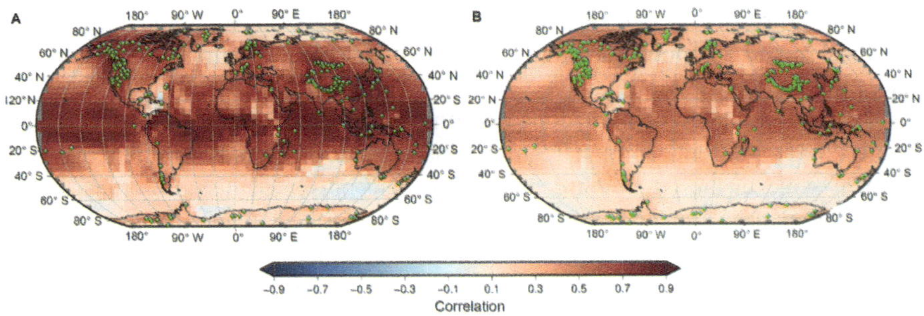


**Fig. 7.9** An example of a pseudo-reconstruction of annual mean temperature using a pseudo-proxy network mimicking the PAGES network of Fig. 7.6. The maps depict the correlations between the target (simulated temperature) over the past 1000 years with the pseudo-reconstructed temperature. (**a**) shows the results using perfect (noise-free) pseudo-proxies, (**b**) shows the results using more realistic pseudo-proxies, in which the local correlation between the pseudo-proxy record and the model grid-cell simulated temperature is uniformly set to 0.5. From Gómez-Navarro et al. (2017)

The idea of using climate models as a synthetic laboratory to test reconstruction methods was further extended to climate reconstructions based on borehole temperature measurements. This *surface temperature proxy* is somewhat special because it is the only *direct* measurement of past temperatures, albeit perturbed by the passage of time. The vertical temperature profile measured in current boreholes—typically 500 meters deep from the Earth's surface—provides information about past surface temperatures. The temperature profile that can be measured today is modulated by the variations of the heat flux from the surface—and thus by the near-surface air temperature—over the past centuries. The surface heat flux requires several centuries to reach those depths and impacts the temperature of those deeper layers (Beltrami et al. 2002). However, the surface heat flux can also be distorted by snow and soil moisture, i.e., climate factors different than air temperature, and also by the geological heat flux from the Earth's interior, so it is not straightforward to translate near-surface temperature profiles into the history of the downward heat flux. Earth System Models represent an ideal laboratory since these models also simulate the state of the upper soil layers, soil moisture and snow cover. The analysis of these simulations allowed us to quantify the impact of the surface climatic factors on the modulation of downward surface-heat flux and thus to estimate the possible errors inherent in temperature reconstructions derived from boreholes (González-Rouco et al. 2003, 2009; Mann et al. 2009; Melo-Aguilar et al. 2020).

# References


Alexandersson H, Schmith T, Iden K, Tuomenvirta H (1998) Long-term trend variations of the storm climate over NW Europe. Global Atmos Ocean Syst 6:97–120

Asena Q (2021) Pseudoproxy modelling to explore ecosystem dynamics and assess uncertainties in palaeoecology. PhD thesis, ResearchSpace@ Auckland

Baker A, Bradley C, Phipps S, Fischer M, Fairchild I, Fuller L, Spötl C, Azcurra C (2012) Millennial-length forward models and pseudoproxies of stalagmite $\delta^{18}$O: an example from NW Scotland. Climate Past 8(4):1153–1167. https://doi.org/10.5194/cp-8-1153-2012

Beltrami H, Smerdon JE, Pollack HN, Huang S (2002) Continental heat gain in the global climate system. Geophys Res Lett 29(8):8–1. https://doi.org/10.1029/2001GL014310

Bond G, Heinrich H, Broecker W, Labeyrie L, McManus J, Andrews J, Huon S, Jantschik R, Clasen S, Simet C et al (1992) Evidence for massive discharges of icebergs into the North Atlantic ocean during the last glacial period. Nature 360(6401):245–249

Broecker W, Bond G, Klas M, Clark E, McManus J (1992) Origin of the northern Atlantic's Heinrich events. Climate Dyn 6:265–273

Bürger G, Cubasch U (2005) Are multiproxy climate reconstructions robust? Geophys Res Lett 32(23). https://doi.org/10.1029/2005GL024155

Carter D, Draper L (1988) Has the north-east Atlantic become rougher? Nature 332:494. https://doi.org/10.1038/332494a0

Christiansen B, Schmith T, Thejll P (2009) A surrogate ensemble study of climate reconstruction methods: Stochasticity and robustness. J Climate 22(4):951–976. https://doi.org/10.1175/2008JCLI2301.1

Gómez-Navarro JJ, Zorita E, Raible CC, Neukom R (2017) Pseudo-proxy tests of the analogue method to reconstruct spatially resolved global temperature during the Common Era. Climate Past 13(6):629–648. https://doi.org/10.5194/cp-13-629-2017

González-Rouco F, von Storch H, Zorita E (2003) Deep soil temperature as proxy for surface air-temperature in a coupled model simulation of the last thousand years. Geophys Res Lett 30(21). https://doi.org/10.1029/2003GL018264

González-Rouco JF, Beltrami H, Zorita E, Stevens M (2009) Borehole climatology: a discussion based on contributions from climate modeling. Climate Past 5(1):97–127. https://doi.org/10.5194/cp-5-97-2009

Hodell D, Crowhurst S, Lourens L, Margari V, Nicolson J, Rolfe JE, Skinner LC, Thomas N, Tzedakis PC, Mleneck-Vautravers MJ et al (2022) A 1.5-million-year record of orbital and millennial climate variability in the North Atlantic. Climate Discuss 2022:1–58

Hogben N (1994) Increases in wave heights over the North Atlantic: A review of the evidence and some implications for the naval architect. Trans Roy Inst Naval Arch W5:93–101

Karl T, Quayle R, Groisman P (1993) Detecting climate variations and change: New challenges for observing and data management systems. J Climate 6:1481–1494

Klein F, Abram NJ, Curran MA, Goosse H, Goursaud S, Masson-Delmotte V, Moy A, Neukom R, Orsi A, Sjolte J et al (2019) Assessing the robustness of Antarctic temperature reconstructions over the past 2 millennia using pseudoproxy and data assimilation experiments. Climate Past 15(2):661–684. https://doi.org/10.5194/cp-15-661-2019

Koslowski G, Glaser R (1999) Variations in reconstructed ice winter severity in the western Baltic from 1501 to 1995, and their implications for the North Atlantic Oscillation. Climatic Change 41(2):175–191. https://doi.org/10.1023/A:1005466226797

Koslowski G, Loewe P (1994) The western Baltic sea ice season in terms of a mass-related severity index: 1879–1992: Part 1. temporal variability and association with the North Atlantic Oscillation. Tellus A 46(1):66–74

Krieger D, Krueger O, Feser F, Weisse R, Tinz B, von Storch H (2021) German Bight storm activity, 1897–2018. Int J Climatol 41:E2159–E2177. https://doi.org/10.1002/joc.6837

Krueger O, von Storch H (2011) Evaluation of an air pressure-based proxy for storm activity. J Climate 24:2612–2619. https://doi.org/10.1175/2011JCLI3913.1

Krueger O, Schenk F, Feser F, Weisse R (2013) Inconsistencies between long-term trends in storminess derived from the 20CR reanalysis and observations. J Climate 26:868–874. https://doi.org/10.1175/JCLI-D-12-00309.1

Krueger O, Feser F, Weisse R (2019) Northeast Atlantic storm activity and its uncertainty from the late nineteenth to the twenty-first century. J Climate 32:1919–1931

Lindenberg J, Mengelkamp HT, Rosenhagen G (2012) Representativity of near surface wind measurements from coastal stations at the German Bight. Meteor Z 21:99–106. https://doi.org/10.1127/0941-2948/2012/0131

Liu J, Shu Y, Zhang L, Xue F, Yang OW (1999) Traffic modeling based on FARIMA models. In: Engineering solutions for the next millennium. 1999 IEEE Canadian conference on electrical and computer engineering (Cat. No. 99TH8411), vol 1. IEEE, pp 162–167. https://doi.org/10.1109/CCECE.1999.807189

Ludescher J, Bunde A, Büntgen U, Schellnhuber HJ (2020) Setting the tree-ring record straight. Climate Dyn 55(11):3017–3024. https://doi.org/10.1007/s00382-020-05433-w

Makse HA, Havlin S, Schwartz M, Stanley HE (1996) Method for generating long-range correlations for large systems. Phys Rev E 53(5):5445. https://doi.org/10.1103/PhysRevE.53.5445

Mann ME, Rutherford S (2002) Climate reconstruction using 'Pseudoproxies'. Geophys Res Lett 29(10):139–1. https://doi.org/10.1029/2001GL014554

Mann ME, Bradley RS, Hughes MK (1998) Global-scale temperature patterns and climate forcing over the past six centuries. Nature 392(6678):779–787. https://doi.org/10.1038/33859

Mann ME, Schmidt GA, Miller SK, LeGrande AN (2009) Potential biases in inferring Holocene temperature trends from long-term borehole information. Geophys Res Lett 36(5). https://doi.org/10.1029/2008GL036354

Matulla C, Schöner W, Alexandersson H, von Storch H, Wang X (2008) European storminess: Late 19th century to present. Climate Dyn 31:125–130. https://doi.org/10.1007/s00382-007-0333-y

Matulla C, Hofstätter M, Auer I, Böhm R, Maugeri M, von Storch H, Krueger O (2012) Storminess in northern Italy and the Adriatic Sea reaching back to 1760. Phys Chem Earth Parts A/B/C 40–41:80–85. https://doi.org/10.1016/j.pce.2011.04.010

Melo-Aguilar C, González-Rouco JF, García-Bustamante E, Steinert N, Jungclaus JH, Navarro J, Roldán-Gómez PJ (2020) Methodological and physical biases in global to subcontinental borehole temperature reconstructions: an assessment from a pseudo-proxy perspective. Climate Past 16(2):453–474. https://doi.org/10.5194/cp-16-453-2020

National Research Council, Division on Earth, Life Studies, Board on Atmospheric Sciences, Committee on Surface Temperature Reconstructions for the Last 2 Y (2007) Surface temperature reconstructions for the last 2,000 years. National Academies Press

Neu HJA (1984) Interannual variations and longer-term changes in the sea state of the North Atlantic from 1970 to 1982. J Geophys Res Oceans 89(C4):6397–6402. https://doi.org/10.1029/JC089iC04p06397

PAGES 2k Consortium (2013) Continental-scale temperature variability during the past two millennia. Nature Geosci 6(5):339–346. https://doi.org/10.1038/NGEO1797

Pielke RA, Landsea CW (1998) Normalized hurricane damages in the United States: 1925–95. Weather Forecast 13:621–631. https://doi.org/10.1175/1520-0434(1998)013<0621:NHDITU>2.0.CO;2

Rutherford S, Mann M, Osborn T, Briffa K, Jones PD, Bradley R, Hughes M (2005) Proxy-based Northern Hemisphere surface temperature reconstructions: Sensitivity to method, predictor network, target season, and target domain. J Climate 18(13):2308–2329. https://doi.org/10.1175/JCLI3351.1

Schmelzer N, Holfort J (2011) Ice winter severity in the western Baltic Sea in the period of 1301–1500: comparison with other relevant data. Int J Climatol 31(7):1094–1098. https://doi.org/10.1002/joc.2337

Schmidt H, von Storch H (1993) German Bight storms analysed. Nature 365:791. https://doi.org/10.1038/365791a0

Smerdon JE (2012) Climate models as a test bed for climate reconstruction methods: pseudoproxy experiments. Wiley Interdiscip Rev Climate Change 3(1):63–77. https://doi.org/10.1002/wcc.149

Sturm C, Zhang Q, Noone D (2010) An introduction to stable water isotopes in climate models: benefits of forward proxy modelling for paleoclimatology. Climate Past 6(1):115–129. https://doi.org/10.5194/cp-6-115-2010

Tingley MP, Huybers P (2010) A Bayesian algorithm for reconstructing climate anomalies in space and time. Part I: Development and applications to paleoclimate reconstruction problems. J Climate 23(10):2759–2781. https://doi.org/10.1175/2009JCLI3015.1

Tolwinski-Ward SE, Evans MN, Hughes MK, Anchukaitis KJ (2011) An efficient forward model of the climate controls on interannual variation in tree-ring width. Climate Dyn 36(11):2419–2439. https://doi.org/10.1007/s00382-010-0945-5

von Storch H, Emeis K (2017) Hartmut Heinrich – The unknown world-famous climate researcher of Hamburg. An interview. https://doi.org/10.13140/RG.2.2.22909.15846

von Storch H, Guddal J, Iden KA, Jonsson T, Perlwitz J, Reistad M, de Ronde J, Schmidt H, Zorita E (1993) Changing statistics of storms in the North Atlantic? MPI report 116, Max-Planck-Institut für Meteorologie

von Storch H, Zorita E, Jones JM, González-Rouco YDF, Tett SF (2004) Reconstructing past climate from noisy data. Science 306(5696):679–682

von Storch H, Zorita E, González-Rouco F (2009) Assessment of three temperature reconstruction methods in the virtual reality of a climate simulation. Int J Earth Sci 98(1):67–82. https://doi.org/10.1007/s00531-008-0349-5

Wang XL, Wan H, Zwiers FW, Swail VR, Compo GP, Allan RJ, Vose RS, Jourdain S, Yin X (2011) Trends and low-frequency variability of storminess over western Europe, 1878–2007. Climate Dyn 37:2355–2371

WASA group (1998) Changing waves and storms in the Northeast Atlantic? Bull Am Meteorol Soc 79:741–760. https://doi.org/10.1175/1520-0477(1998)079<0741:CWASIT>2.0.CO;2

# Index





H. von Storch et al., *Climate Science Concepts Born in Hamburg*,
https://doi.org/10.1007/978-3-031-81108-1

The manufacturer's authorised representative in the EU is Springer Nature Customer Service Centre GmbH, Europaplatz 3, 69115 Heidelberg, Germany. If you have any concerns regarding our products, please contact ProductSafety@springernature.com

Printed and bound by CPI Group (UK) Ltd, Croydon, CR0 4YY
30/09/2026
02225254-0019